HODGSON'S ESTIMATOR
AND
CONTRACTOR'S GUIDE
for PRICING BUILDER'S WORK

Describing Reliable Methods of Pricing Builder's Quantities for Competitive Work, showing in brief and concise form the methods generally employed by the most successful contractors.

GIVING FULL DETAILS FOR ESTIMATING COST

BY COST PER CUBIC FOOT OF SIMILAR BUILDINGS ESTIMATING BY THE SQUARE, ESTIMATING IN ROUGH QUANTITIES, ESTIMATING PER UNIT OF ACCOMMODATION, ESTIMATING BY ACCURATE QUANTITIES, ETC.

With Many Tables, Rules *and* Useful Memoranda

BY FRED T. HODGSON
MEMBER OF O. A. A.
Author of "Treatise on Uses of The Steel Square," "Modern Carpentry," "Architectural Drawing Self-taught;" "Hodgson's Wood-worker and Up-to-date Hardwood Finisher," etc.

CHICAGO
FREDERICK J. DRAKE & CO., PUBLISHERS
1904

PREFACE

The man who undertakes to build a book worth anything on estimating the cost of proposed buildings, is "up against" a pretty serious proposition. Not that such a book cannot be written that will be of great use to builders, but because of the ever shifting of prices of labor and materials, and the constant changing of methods and appliances. Figures that may be all right and correct for the work of to-day, may be entirely wrong and misleading to-morrow, and this is the main objection to works on estimating.

There are, however, certain rules and constants of measurements the estimator may employ when figuring up the cost of proposed buildings that may be relied upon as being correct, and in this work I have endeavored to show these rules and constants in as clear and understandable a manner as I know how, and I think my efforts have not been failures.

After all, the main factor to be employed in the make-up of an estimate is experienced judgment. No matter how much mathematics a man may be master of, if he has not experience in building matters and mature judgment to guide him, he can never become a reliable estimator. A good judgment may be born in a man, but experience can be gained only by a certain amount of labor and drudgery. As in other departments of science and art, there "is no royal road" to estimating, unless it be that which leads to guess work and financial disaster; therefore, let me press, at the outset, on the minds of all owners of this work that an expert estimator can only become so by study and by a mastery of all the details that enter into the make-up of a building, added to a keen judgment and a comprehensive knowledge of the markets, labor, and materials employed on the proposed works.

Many an honest builder, good mechanic, and clever constructionist, has come to grief by taking contracts too low, because of his lack of knowledge in estimating, and thus not only does himself a great wrong, but he also disorganizes the whole building business in his neighborhood; for if he undertakes to do a certain job for a given price, his neighbors will expect to get similar work done for similar prices, and rival contractors then strain every nerve to get their estimates down to his level, and in doing so inferior materials are used, "scamping" is resorted to, and labor is crushed and cheapened to meet the conditions.

In the following work I have made an endeavor to place within the reach of every workman of experience an opportunity of qualifying himself to undertake the preparing of figures for work, so that he can make his tenders within the limits of reason—not absurdly high, or ridiculously low—so that only with a pen or pencil and this book he may be able to figure out and price a set of quantities in short order.

Great pains have been taken to collect such exact information as may be found useful in estimating, either in the office or on the building, with the object of forming what I believe will prove a valuable addition to building literature in other directions than that of simply being a price book.

During the last few years, materials of all kinds, raw or finished, have risen in price from 25 to 50 per cent, and labor has gone up in nearly the same proportion, and the end is not yet, and artificial values have been created, and this continual fluctuation must always be considered when estimating, because no rules can be so devised as to be as elastic as prices and material men's quotations. This fact, or facts, only go to show that wherever prices are given in books of this sort, they should be accepted with salt. Notwithstanding this, however, the principles of estimating, as herein set forth, still hold good in so far as quantities and methods are concerned.

Collingwood, Ontario, May 1, 1904.

FRED T. HODGSON.

HODGSON'S ESTIMATOR AND CONTRACTOR'S GUIDE

INTRODUCTORY

Estimating the cost of a proposed building of any kind is not of a nature to attract the young workman, as it is a dull, dry, and methodical business and only the requirements of a sordid and money-making necessity compel the builder to wade through mazes of figures to attain the desired result.

If the writer had consulted his own pleasure and followed his inclinations he would not have written at all, or on a subject more congenial to his taste; but from long experience and observation and more or less practice, he has witnessed so much ignorance and inaptitude on the part of young men who have essayed to be builders and contractors that, with the advice of his publishers, he has undertaken to prepare this work on estimating, because it has been thought that a work of the kind may prove useful and of benefit to the young man who aspires to be a master builder or a contractor, and who *may*, if he chooses to go to the trouble, make himself fairly competent to arrive at the cost of any reasonable sized building. It may as well be understood at the outset, however, that there is no royal road by which eminence as an estimator can be attained. No matter what system or method may be adopted, correctness can only be reached through an avenue of labor and sound judgment. The best and most ingenious writers on the subject of estimating have never yet been able to discover or devise a method where the cost of a building may be "jumped at'' at first sight. The system of cubing is, perhaps, the easiest of all methods, but is not a system the experienced builder would care to follow altogether, unless a large margin of profits and contingencies are provided for.

While it will be impossible for me to so prepare this work as to be as entertaining as a novel, I will, to the best of my abilities, make it as easy to understand by the every-day workman as it possibly can be.

Estimating is the most difficult task the builder has to deal with, and too much care cannot be taken, even if the quantities are supplied, if a correct tender is wanted. Many who tender make up their prices in a haphazard manner, often depending on trade catalogues, price lists or newspaper quotations for data, using their judgment, whether experienced or not, and without a full or even a fair knowledge of the scientific methods which underlie the proper formulating of a true estimate. Prices which enable successful contractors to calculate values for themselves are obtained by dissecting, taking asunder and examining the various elements that go to make them up, the complete result being shown in a final bill of quantities, labor and other costs.

It will be impossible to make this work a mere hand-book of builders' prices, as what may be the ruling price of labor or material to-day may be very much different to-morrow, as in these days of continual change there can be no such thing as "constants" in prices. I can give quantities, however, and describe the proper methods of obtaining them, and can convey to the student the principles upon which correct estimating is based, and offer here and there the prices of labor as *now current* in the larger cities, not to be followed, but simply to give an idea of the cost of various kinds of work when no other data is available.

No man can be a successful contractor who does not attend strictly to his bookkeeping, so that he can tell in a moment, by reference to his books, the exact amount of profit or loss on the various jobs of work he has completed. This is important, inasmuch as the mistakes in estimating may be traced to their source, and thus be avoided when similar jobs are being figured on; and much trouble and disappointment may be avoided by having the accounts on every job itemized and kept in proper order. I will have more to say on this subject later on.

All estimates should be retained, properly labeled, and put in some place where they can be found when required, whether the work for which they were prepared is secured or not, for they will often prove of great service for future reference; and the estimator should make a note of each particular on which he may have priced too high or too low, if his tender is not accepted. If the work is secured, the cost of each particular item in the building should be compared with the estimated price, and a note should be made on the margin of the original estimate showing the discrepancy, if any, between estimated and actual price. A correct account of all labor, how employed, should also be kept, so that the contractor may know from *actual facts* exactly what a piece of work costs, or the number of days or hours it required to perform such and such work, also amount and cost of materials on the same work; then, in preparing other estimates, he will have something tangible to base his figures on. It is better to estimate on days or hours for time, and on quantities for materials, because of the continual fluctuations in price of labor and materials of all kinds. If it takes $2\frac{1}{2}$ days, of 9 hours each, to execute a piece of work, the figuring on this is quite simple, for all we have to do is to multiply the number of hours by the price per hour for labor; suppose this to be 30 cents an hour, then we have $22\frac{1}{2} \times 30 = 675$. That is, in $2\frac{1}{2}$ days, at 9 hours per day, we get $22\frac{1}{2}$ hours at 30 cents per hour, which will make 675 cents, or $6\frac{3}{4}$ dollars. Quantities may be figured in a similar manner. If the work requires 150 feet of material, then charge that at current rates, whatever these may be; then add cost of labor and material together, and you have the bare cost of the work. To these, of course, must be added cartage, profit and any other materials that have been employed on the work, such as nails, screws, glue, paint, or anything else. By following this course, a record of all work done and estimated for will always be at hand, and it is surprising how much the labor of estimating may be reduced by a strict adherence to this system, as a comparison with work done and work to do may be made in a few minutes, and the difference in prices of labor then and now adjusted so that no loss will occur to the contractor.

The variations in tenders for the same work are often surprising. I have seen estimates, particularly in carpenter's and joiner's work, run up to as much as *50 per cent* above the tenders of competing contractors, yet the lowest bidder made money. Competent estimators never make such wide errors as this, though often they do not keep close to the wind; and while mistakes will continue to be made, even by the very best estimators, by omissions, "doubling up," and using wrong dimensions, the mistakes may be narrowed down to a very small area if system, care and judgment be exercised when the estimates are made. It will be the object of this little volume to narrow this area of error to the smallest possible limits, and to show the estimator how to avoid grievous errors and make his estimates more satisfactory and reliable.

CATALOGUES AND PRICE LISTS

No builder's office can be well equipped except it contains the latest catalogues and price lists available, for on these the estimator must, to some extent, be dependent in his figuring on the cost of most of the material that goes in the work. Bricks, stone, lime, cement, lumber, hardware, and factory-made stuff may have their prices approximated from these publications, but the shrewd estimator, while making use of these aids, does not rely upon them for serious pricing. They help considerably, as they contain a lot of condensed information regarding prices and building; but they are not always to be depended upon, as they are not always compiled in a scientific way. For example, some of the prices include trade discount, some do not, while others are merely the ordinary list prices of merchants' catalogues. The discount in itself largely varies, and there are two, and often more, discounts--a trade discount and a cash discount—and other mysterious discounts, such as 30% and 5%, which means 30 per cent off and 5 per cent off the balance; and again, the percentages are not uniform; one merchant may have one discount, another another, so in all cases it is best to get prices and discounts direct from the merchant wherever possible. The diversities in discounts are innumerable, and it is the estimator's duty to get definite information as to prices and discounts as prevailing in the locality where the work is to be done.

Builders' prices are broadly made up of two divisions, labor and material, to which may be added a third, namely, profit. The cost of labor and material vary from time to time, and from place to place, and do not fluctuate similarly. Sometimes labor may be high and materials low in price, and at other times materials may be high and labor low, so that no given rules can be formed to meet these conditions and be constant, and this fact rules price-books out of the race of accuracy for any length of time. Such things as closeness or slackness of supervision, misunderstandings as to quality of workmanship or materials, worrying by the architect, delay in furnishing detail drawings, differences in locality and site, frost and bad weather, sudden and unexpected rises and falls in the market, etc., will all help to alter the conditions of profit or loss for the contractor, and the extent of which is almost impossible to measure.

When, however, the contractor has worked out a series of prices for himself, to suit both time and locality, he must be on the alert for parallel cases to avoid the great labor involved in making calculations afresh every time a new estimate is required. In fact, he should carefully prepare a sort of price-book for himself, suitable to the conditions, and so arranged that it can be revised from time to time. Thus a consistency in pricing would result, which is of considerable importance.

As already stated, the builder will be confronted with several grades of discounts, and among them will be a *cash* discount. This may be more or less or anywhere within the limits of from 2 to 5 per cent, and it should be the aim of the contractor to get the best discounts to be had, providing the materials or goods are up to the standard demanded by the specifications. Sometimes it may happen that on special goods or some particular make of hardware or other items, no discounts are allowed. This, however, can only happen when a dealer has the sole control of these special goods, or when there is a scarcity of them in the market, or when a sudden demand for them arises. These conditions, however, seldom or never occur, so they may hardly be considered. In the practice of a shrewd contractor, the question of discounts enters largely into the make-up of an estimate, particularly where close competition is likely to be met with.

The question of profit is one that must be well considered when estimating; 10 per cent is the least amount a builder can accept, exclusive of established charges, and this should be added to each individual price, and no provision should, under ordinary conditions, be made for any trade discounts, as these are expected to swell the profits. Some estimators when pricing bills of quantities prefer to add a lump sum as profit at the end of a bill. This, however, is not a good thing to do, as it gives no correct method of knowing what the profits are.

For wood or materials on small jobs, where both are limited, the profit should be higher, as the total expenditure in such a case is much more in proportion; therefore the percentage of profit should never be less than 15 per cent on work costing up to $2,500, but above this amount a smaller percentage would perhaps be sufficient.

The large contractor, who may perhaps own his own brickyard, quarry or factory, with extensive premises and rapid-working labor appliances of all kinds, can naturally turn out work cheaper and more expeditiously and at a greater profit to himself than the small contractor who possesses none of these appliances and aids. Often the latter, in order to save himself from loss, is obliged to scamp the work and use inferior materials, which he can frequently "get in" without the architect being able to detect it; he is often obliged to do this in order to keep himself afloat. My advice in cases of this kind is, that the lower contractor should confine himself to certain prices—that will pay him—and if he cannot win the work for these prices he had better leave the work for the larger contractors, and thereby preserve his reputation and his money. The small contractor can always find plenty of work to do if he but gets a good name for doing his work well and according to specification.

Where there are dozens of doors made from one pattern, as many window frames and sashes, and hundreds of feet of mouldings in wood or stone of one shape and size, they can be rattled out by machinery in short order and at a comparatively low cost, and this is an item the estimator must consider, as it will aid materially in keeping down the total amount of tender; in any case, however, experience and judgment in such matters are required before a definite amount can be decided upon.

With reference to terms of payment, it is always better that the contractor gets his money often, as it enables him to push his work with greater vigor, and gives him a chance of making the best cash discounts when purchasing materials, and, on these several accounts, he will be able to make a lower bid for the work than otherwise. The reserve to be deducted from each payment should never exceed 25 per cent, which is considered ample to cover any liens of workmen or material men and safeguard the interests of the owner. There are certain fixed charges or provisions in contracting that must not be overlooked. These consist of salaries, depreciation of plant, tools, machinery, rent of premises, lights, water, and interest in capital invested, of which the new work must pay its proportional share, and these charges should be kept separate and added to the estimate along with the percentage of profit. Such charges are commonly placed at 6 per cent interest on capital invested, and 3 per cent for depreciation of plant, etc. Sometimes they are classed in two categories: 6 per cent on work done on the building, and 8 per cent on work done in the contractor's factory or shops. These percentages, however, are somewhat arbitrary, and should be the result rather of experience and good judgment than any fixed rules, and the foregoing remarks are offered rather as reminders that some allowances must be made for each item when estimating, otherwise they might be overlooked.

The question of transportation is one also that enters largely into the cost of work. If the works are situated nearby the office and establishment of the contractor, the question will not be so formidable as when the work is some distance away, as the greater part of the material will very likely be near the ground and may only require handling and teaming once; but where the work is at a distance, the expense of getting the material on the ground will necessarily be much greater. When conditions will admit of it, it is always better and cheaper to have material shipped by boat than by rail, or long hauls by team, and the estimator should make himself familiar with all the ways of communication to the spot where the building or buildings are to be erected, and should get a schedule of rates from all the lines running to that point. A good idea is to get a map of the district which shows all the railway and water communication; then the shortest and best routes can be chosen, providing the rates are satisfactory. As I stated before, it is much better, when it can be done, to ship by water than by land, as because of the absence of vibration, fine work will be less likely to be injured or scratched during transit, and, as a rule, rates are always lower by water than by land. The average rate for the shipment of goods in this country is about 1¾ cents per mile for short hauls, and something less for long hauls.* Rates, however, vary with the different roads and at different times, the highest rates being in winter, in the north, when the waterways are frozen up. Classification, also, has something to do with regulating rates. All goods should be insured or shipped at the carrier's risk, then losses or damages will be covered. If goods have to be packed, or put up in crates or boxes, at least 15 per cent should be allowed for this work and material, and should be charged on the special goods boxed or crated only, but added to the estimates.

Goods sent at carrier's risk that get damaged, should be returned by the same carriers free of cost, and when repaired or renewed should be delivered at the point where first destined, at the cost only of the first shipment of the same goods. That is, the shipper should pay for one shipment only.

Where a quantity of goods of a similar kind is required, a special quotation should be given the con-

---

*Per ton.

tractor by the dealer, and this should never be overlooked, for it is not likely that it will be given if not applied for.

Trade discounts, as a rule, are not publicly stated in trade catalogues or circulars; they can be obtained only on private application. Their amounts greatly depend on the quantity of goods ordered, and the larger the order the larger the percentage given.

The foregoing remarks are offered as a sort of preliminary and should be well considered by the intending estimator, as they contain much that will tend to smooth the way towards accuracy in making up a tender, and, if followed attentively, will enable the estimator, along with the rules that follow, to get at a result that will be nearly correct and satisfactory.

SYSTEM IN ESTIMATING

The estimator should follow some well-defined system in his work, in order that he may know he has not overlooked anything, for one of the dangers is that of omission. To overlook the roof—as I have known one instance of the kind—the floors, the doors, or anything else, is a serious matter, and in order to prevent this as much as possible I have prepared a list of items which I give further on, and which may be called a "Tickler" or a "Reminder" of what will be required to consider when making an estimate of a building complete.

When erecting a structure of any kind, work should commence at the earth, so the first thing estimated, following the same rule, should be the excavations for cellar, drainage, foundations, trenches, and other similar work, then the preparing and the laying of the foundations, whether of stone, concrete, or brick; and the same order should be followed throughout the whole building, until the whole is fully completed, from turning the first shovelful of earth until the last piece of finished work is put in place.

The following items will remind the estimator of the things to be figured on as he works his way upwards:

Inspection of site
Examination of soil
Note if gravel, soil, or sand
Figure accordingly
Get number of cubic yards
The distance to be removed
Where to be deposited
Pumping water
How drained
Sewerage
What depth of drains
Depth of cellar
Depth of foundation walls
Width of footings
Rock blasting
Shoring banks
Piling for foundations
Sheet piling
Excavations for piers
Cesspool
Cistern
Trenches
Cuttings for water pipes
Grading
Leveling cellar floor
W. C. for workmen
Removing fences
Grubbing out tree stumps
Removing surplus soil
Removing debris
Sodding
Carriageways
Footpaths
Driveways to rear
Tamping earth
Concreting foundation
Openings for drain pipes
Laying drain pipes
Area of all tiles
Weeping tiles
Elbows and bends
Traps of all kinds
Intake water pipes
Waste pipes
Footings
Cellar walls
Furnace room
Walls laid in cement
Walls laid in lime mortar
Walls built up of concrete
Stone walls, field stone
Stone walls, quarried stone
Stone walls, dimension stone
Brick walls for cellar
Amount of stone
Amount of bricks
Amount of concrete
Cellar steps
Cellar windows
Cellar doors
Cellar partitions
Cellar coping stones
Cellar sills and lintels
Bond stones

Cellar water closet
Water taps, etc.
Concrete and cement floor
Plank floor
Earth floor tamped
Wine cellar
Vegetable cellar
Coal storage bins
Coal chute
Ashes receiver
Cellar stairs
Preserve closet
Shelving
Plastering walls and ceilings
Damp courses in walls
Double sashes in windows
Doors, what kind
Fireplace and chimney
Laundry tubs
Hot and cold water supply
Furnace and attachments
Furnace, hot water
Furnace, steam water
Furnace, hot air
Gas jets, how many
Electric lights, how many
Laundry table
Clothes drying device
Mangle
Chimney piece
Stove rings
Registers
Cellar finish
Wardrobe hooks and pins
Cupboards and drawers
Tool room
Wash bowl and stand
Kind of hardware
Ground floor
Number of rooms
Number of doors
Number of windows
Style of doors
Style of windows
Sizes of doors and windows
Thickness of doors and windows
Kind of glass
How windows are hung
Hardwood or pine finish
Outside walls, stone, brick or wood
Thickness of walls
If stone, rock face
Tooled, rubbed
Cross tooth chiseled
Crandalled
Brick wall
Thickness of brick walls
Common bricks
Pressed bricks
First, second or third quality
Mixed, brick and stone
Walls ornamented
Walls left plain
Window finish
Urinals
Slate slabs
Exterior window finish
Interior window finish
Exterior door finish
Interior door finish
Betting courses
Sailing courses
Laid in cement or mortar
Front steps, stone
Front steps, cement or wood
Hall entrance
Double floor, pine
Hardwood floor
Parquet floor in some rooms
Tile floors
Dimensions of joists

Thickness of floors
Height of ceilings
Stairs, straight
Stairs, winding
Stairs, platform
Pine or hardwood
Kind of hardwood
Styles of newels and balusters
Plain finish in rooms
Ornamental finish in rooms
Fret and grill work
Arches, plain or otherwise
Styles of plastering
Stucco cornices
Styles of cornices
Sliding doors
Fireplaces
How many
Mantelpieces
Mantelpieces, plain or ornamental
How finished
Other wood finish
Pillars, columns or brackets
Base and plinth
Style of trimmings
Style of hardware
Cost of hardware
Grates and tiles
Mirrors
Gas lighting
Jets and gasoliers
Electric lighting
Electroliers and brackets
Piping for gas
Wiring for electric lights
Fitting clothes closets
Fitting up den
Fitting up closets
Fitting up cellar stairs
Fitting up dining room
Fitting up other rooms
Kitchen finish
Tubs, sinks, dresser
Cupboards, china closet
Butler's pantry
General pantry
Range
Steam cooker
Chimneys
Ventilation
Painting
Varnishing
Wainscot
Panelings
Washstands
Marble facings for walls
Double windows
Sashes, weights and cords
Box frames
Plain frames
Window stools
Inside shutters
Inside blinds
Splay boxes
Tiled hearths
Sash locks
Tiled facings
Back stairs
Servant's room
Bay window
Oriels
Veranda
Front porch
Rear porch
Stoop
Back areas
Front areas
Iron railings
Stone railings
Balconies
Window hoods

Door hoods
Door stops
Door springs
Plate glass
Stained glass
Niches
Closet fittings
Provide for heating
Conservatory
Corrugated glass
Skylights
Handrail, oak or mahogany
Bracketed stairs
Anchors and tie irons
Vaults
Angle irons
Bond timbers
Carving, if any
Scaffolding
Temporary enclosure
Iron beams
Iron columns
Gas pipe pillars
Water on main floor
Taps, nickel plated
Taps, plain
Glazier's work
Meters, syphons
Elbows, pendents
Painting
Paper hanging
Iron pipes
Lead pipes
Brass pipes
Washers, wastes
Plugs, grating
Pumps, suction pipes
Wall hooks, supply pipes
Cast iron work
Wrought iron work
Stucco work generally
Stucco friezes, enrichments
Stucco pateras, panels
Stucco moldings
Stucco beads, straight
Stucco beads over arches
Stucco arrises, quirks
Stucco reveals angles
Stucco centerpieces
General plastering
Two coats
Three coats
Lathing
Quality of laths
Sand, lime and hair
Plaster of Paris
Clean water
Sound story joists
Studding for partitions
Beams
Trimmers for hearths
Trimmers for stairs
Trimmers for chimneys
Strapping walls
Dimensions of strapping
Wooden bricks
Plugging walls
Nailing strips
Temporary sashes
Lanterns
Louvres
Thresholds
If metal ceilings
If metal cornices
Metal centerpieces
Bridging joists
Bridging studding
Dimensions of studs
Double partitions for sliding doors
Lining pocket of sliding doors
Hanging sliding doors

Framing wooden house
Boarding inside
Boarding outside
Boarding both sides
Papering one or both sides
Horizontal boarding
Diagonal boarding
Tar paper or plain paper
Outriggers
Towers
Two-story bay windows
Two-story oriels
Two-story balcony
Two-story porches
Two-story verandas
Three or more stories of same
Iron railings for balconies
Wood railings for same
Ornamental iron column
Ornamental brackets, iron
Iron supports for platform
Iron trusses for balconies
Iron plates for piers
Other iron work
Siding frame buildings
Half-timbered building
Rough cast building
Brick veneered building
Wood cornice outside
Metal cornice outside
Shingle cornice outside
Brick cornice outside
Stone cornice outside
Attic floor joists
Rafters
Collar beams
Trusses for roofs
Framing for dormers
Framing for eye-winkers
Dormer windows
Chimney stacks
Framing roof
Boarding roof
Mortar under shingles
Mortar under slate
Asbestos paper under covering
Common paper under covering
Shingle roof
Slate roof
Tile roof
Composition roof
Tin roof
Galvanized iron roof
Roofs painted
Flashing of all kinds
Tin flashings
Zinc flashings
Galvanized iron flashings
Eave troughs
Conductor pipes
Size of conductor pipes
Mansard roof
Saddle roof
Hip roof
Flat roof
Tower roof
Square tower roof
Conical roof
Steeple roof
Polygon roof
Bay window roof
Porch roof
Roof over balcony
Veranda roof
Framings for veranda
Chamber floors
Attic floors
Bedroom fittings
Number of doors in bedrooms
Washbasins
Closets, Drawers and fitments
Servants' bedrooms

Hall, sewing room
Continuous stairway
Bathroom and fitments
Water closet, in what style
Bathroom washstand
Linen closet
Nursery
Fireplaces
Mantels
Tiling for fireplaces
Base, style of finish
Built in seats
Finish in main bedroom
Finish in nursery
Finish in servant's room
Finish in bathroom
Finish in hall
Finish in closets
Openings and arches
Style of painting
Pine finish
Hardwood finish
Character of finish
Cost of hardware
Style and cost of bath tub
Style of water closet
Marble washstand
Tiled walls
Tiled floor
Marble lined walls
Ventilation
Air ducts
Register
Bath trimmings
Shower bath
Hot and cold water
Stairway to attic
Attic storerooms
Attic, clothes drying room
Children's playroom in attic
Inside trim of dormer windows
General finish of attic
Water closet and lavatory in attic
Painting in attic
Attic doors
Heating attic
Attic storeroom
Children's toy room
Hall in attic
Railing around attic stairway
Closets in attic
Water in attic
Plastering in attic
Attic walls all boarded
Matched ceiling in attic
Attic hardware
Chimney tops
Style of chimney tops
Chimney pots
Finishing top of chimney
Stone tops
Cement tops
Metal tops
Roof decks
Railing for decks
Rolls for ridges
Cresting for ridges
Wood cresting
Metal crestings
Terra cotta crestings
Terra cotta panels
Terra cotta work generally
Hatchway in deck
Scuttle in deck
Lead work
Copper work
Tin work
Roof painting
Painted or dipped shingles
Stairs to roof or deck
Flagpole

Halyards
Wire guards
Snow guards
Storm sashes
Storm doors
Screen doors
Wire screens for windows
Wood gables
Brick or stone gables
Half-timbered gables
Plastered gables
Shingled gables
Deafening floors
Deafening walls
Pugging floors
Sub-floors
Diagonal floors
Rough floors
Cellar sleepers
Cedar posts
Chestnut posts
Spandid panels
Lattice work
Entrance approach
Porte-cochère
Stepladders
Refrigerator
Cold storage shelving
Wine bottle racks
Folding partitions
Boxed shutters
Boxed blinds
Sliding blinds
Rolling blinds
Venetian blinds
Dumb waiter
Transom doors
Transom windows
Mullion windows
Circular top windows
Elliptical windows
Double-hung windows
Single-hung windows
Windows, plain
Windows, ornamental
Pavements
Slop hoppers
Vestibule
Vestibule partition
Vestibule floor
Hardwood or tile
Wainscot in vestibule
Wainscot up stairway
Paneled stair strings
Hardwood stairs
Wood-shed
Coal-shed

While the foregoing does not pretend to give all the items that may be required, it offers to the estimator some hints as to what is required, in a general way, for domestic buildings. For factories, stables, barns, warehouses, public buildings, churches, schools, railway stations, and similar work, a more elaborate list would be required, but the estimator should be able to find all the items in the specifications prepared for the work under consideration, and if he is thorough he will add to the list as given above such items, with their cost, as he goes over them when figuring.

DIFFERENT METHODS OF ESTIMATING

It is said there are not less than five different methods of estimating. Four of these are uncertain, but answer for the purpose of getting an approximate cost of some proposed work, and are chiefly made use of by architects and engineers to give their clients an idea of cost before going into actual building operations. The fifth method, which is the only reliable method, is the taking out of exact quantities item by item.

The first of these methods is the estimating by the cost per cubic foot of similar buildings. It is the best known method, and most usually adopted because of its general convenience. The dimensions are best taken by measuring the length and breadth from out to out of walls, and the height from half foundation to half-way up roof. The cubic contents, then obtained, are multiplied by the price per foot cube of some similar building. Sometimes the height is measured from the bottom of footings to half-way up the roof. Cheaper attached structures, such as annexed stables, sheds, etc., should be kept separate and priced lower; while more ornamental portions, like towers and porches, should be valued at a higher rate than the main block. Small buildings cost more in proportion than large ones of the same type.

This cubing system is open to some objections. The lumping together of solids and voids at one rate is certainly not scientific, for the same class of buildings may be divided into many rooms with numerous internal solids in the shape of walls, etc., between;

while another may have comparatively few chambers, creating much empty space. In fact, the proportion of voids to the solid structure is not a fixed quantity, so that the price per cubic foot can never be exactly regulated. This method requires a large experience and a nicety in pricing which the estimator cannot always possess. The description and quality of materials and workmanship, too, are seldom the same; neither are the conditions of contract, and these variations are frequently overlooked when a certain rate per cubic foot is assumed.

A second method is to take out rough quantities and price the items as the estimate proceeds. In this case the quantities of materials and workmanship are ascertained from the drawings in a broad and comprehensive manner, the work being concentrated as much as possible into a few specific items and afterwards priced accordingly. Although this course is perhaps less generally used than any other for estimating purposes, yet it is one of the most reliable methods that can be adopted when time and circumstances do not admit of detailed quantities and prices. The fact that such a method is not more frequently used is probably due in a great measure to the want of a readily accessible table of prices for the different groups of materials and labor. Slightly more time is also required for this purpose than when the cost is arrived at by the cubic contents or any other methods except by detail pricing. The final result, however, is nearer the truth than it would be by cubing. In estimating by this method it will be well to add 10 per cent for contingencies.

When rough quantities are being taken for an approximate estimate, it is desirable that the various descriptions of materials and workmanship should be grouped together so as to form as few separate items as possible; also, in all cases where it can be done, the items should be priced as per square of 100 feet superficial, for the sake of uniformity and convenience.

The walls should be classed according to their materials and thickness, at the same time stating whether external or internal. Each item should include all necessary digging, footings, doors, windows, and finishings of wall surfaces, such as plastering, facings to external walls, etc., so that the item, and consequently the price, shall be inclusive of everything that appertains to the various enclosures or divisions of the building. For this purpose the superficial area of the walls should be obtained by taking the extreme length of each wall by the height from the bottom of the footings to the top of the eaves, in cases where the thickness of the wall is the same throughout. Should the wall vary in thickness, either in its length or height, each portion should be measured separately. No deductions must be made for door, window or other openings. Bay windows, chimneys and other additions of a like nature should be numbered and priced according to their materials and workmanship.

The floors may be dealt with in a manner similar to that described for the walls. The ground and upper floors must be kept separate, and classed according to the materials and finishings required. The item for wood floors on the ground floor to include sleepers, dwarf walls, joists, boarding, hearths, etc., together with a layer of concrete on brick rubbish over the whole area, and all necessary digging for same. Similarly, concrete or other floors will include all materials, labor, and finished surfaces that may be required. The upper floors to be treated in a similar manner.

The item to include all joists, boarding, hearths, ceilings, cornices, and whitening or coloring the same. The roof coverings to be measured on the slope, the item being inclusive of roof trusses, rafters, boarding, shingling, slating or other covering, leadwork, eave-gutters, down pipes, etc. Ceiling joists, ceilings and whitening or coloring to ceilings will also be included in the same items here required.

Drains, gas and water mains, electric wiring, and items of a similar nature, should be taken at per foot or per yard run, according to sizes, including all necessary digging, laying, filling, and removal of surplus materials. Manholes, disconnecting pits, etc., to be numbered and priced according to size and average depth.

Staircases to be taken at per step, or per foot in height, classed according to their widths, and the nature of the materials and finishings. Gas and water fittings to be priced at per light or per tap, including all service-pieces from mains, digging, etc.

Fitments or furnishings generally, such as cupboards, baths, sink, w. c.'s, ranges, grates, mantels, etc., are numbered and priced according to the class of fitments, material and finishings required.

A series of average items and approximate prices adapted to this method of estimating, may be found in this work in some of the tables, rules and memoranda that follow.

The third method of estimating is by the square of 100 feet, which, under some circumstances, is quite convenient for obtaining approximate cost. Its use is principally confined to one-story buildings, such as sheds, stores, schools, churches, chapels, stables, railway stations, bungalows and similar buildings. It may, however, be used for buildings two or more stories in height; but a considerable amount of discrimination and care must be exercised in order that the final result may be relied upon.

The superficial area is obtained by taking the dimensions from out to out of walls at the ground level, so as to include any projection of the plinth or other offset which frequently occurs at the base of a building. The result is commonly called the *plinth area* of the building. Where the materials, workmanship, or height of building or floor varies, each description or height must be kept distinct in order that they may be separately priced.

In case of one-story buildings, the price per square includes foundations, walls, floor, roof, and all finishings. Occasionally data is at hand by which buildings comprising two or more stories, such as warehouses, etc., may be priced in the same way, the price per square of "plinth area" including foundations, walls, ground and upper floors, roof, etc., all complete.

For general purposes, however, it is more convenient to separate the different floors of buildings of more than one story in height and price each floor accordingly.

When this course is adopted for two or more stories, the ground floor is taken to include foundations, floor, walls, ceiling, and all finishings. Upper floor includes floor-joists, flooring, walls, ceilings, finishings, etc., whilst the top floor includes the roof covering in addition.

Sometimes two-story buildings have both floors priced all the same rate, as it is found that the average cost of the ground floor, including the foundations, is about the same as that of the first floor, which includes the roof covering.

It is also useful to remember that the floor area of a certain description of buildings affords some indication of the amount of accommodation provided. For class rooms in schools, the floor area accommodates from seven to ten scholars per square, being an allowance of fourteen to ten superficial feet per child.

Ordinary churches accommodate from nine to twelve persons per square, corresponding to a total floor area of eleven to eight feet superficial per sitting respectively. In mission churches, etc., the floor space frequently averages about seven feet per sitting, or at the rate of fourteen persons per square. These figures include the floor area which is necessarily absorbed by aisles, pulpit, choir, vestry, sanctuary, etc.

The actual amount of floor space required per person for seating accommodation in churches is from 4½ feet to 5½ feet, superficial.

Pews, or sittings, in churches are usually spaced from 34 to 36 inches apart (measuring from back to back of seats), whilst the average length of seat required per person is from 20 to 22 inches.

A fourth method of estimating is by unit of accommodation, and in practice it is found that for certain descriptions of buildings or works, constructed under normal conditions, the cost of such buildings or works varies (within certain limits) in a direct ratio to some known unit of accommodation or requirements.

For such buildings as hospitals, schools, churches, factories, etc., the cost can be approximately given, if the number of patients, children, etc., required to be accommodated is known. On occasions when time will not admit of even a sketch of the proposal being made, this method affords oftentimes the only ready means of ascertaining the approximate cost. Similarly, for certain minor accessories where the cost of materials and construction varies but slightly for units of the same class, as in a range of latrines, etc., the approximate cost can be easily determined in the same way. Data for this method of estimating will be found in the rules I give in this work.

The fifth, and most correct, method of estimating is by taking out accurate quantities of materials and items of all kinds and pricing them as the figures are obtained, and then adding the cost of labor to each item. This may be called a "detailed bill of quantities." This method, because of its entailing so much labor, should be adopted only when it is intended to carry out the work and when a tender is sent in or submitted for work about to be gone on with. It is very laborious, and necessitates great skill and a thorough knowledge of building construction, and particularly of the work to be tendered for, so that the subject is somewhat difficult for young hands to deal with. The system should be divided into three parts or processes, namely, "Taking off," "Abstracting," and "Billing," the last portion showing the prices. In this method a full set of drawings of the work and copious specifications are necessary, so that the estimator can take the dimensions from one and quality of material and character of work from the other. The cost of the various descriptions of material and workmanship are then priced in accordance with the current rates obtained in the locality where the work is to be carried out. This method takes time and much labor, but it has the advantage of being correct, or nearly so, if the work is honestly and faithfully performed. In fact, it is the only method a young contractor should use when commencing business. After years of experience and observation as a builder and contractor, cubing, or one or other of the quick methods, *may* be made use of under certain conditions, where the contractor knows what he is about. My advice, however, is to stick to the old and reliable method of estimating by items. It takes time, but the time and labor are well invested.

The young estimator must necessarily have a fair knowledge of arithmetic, particularly that branch of it termed mensuration, before he can hope to become an expert; indeed, it will be impossible for him to become an expert unless he is good at figures and has some knowledge of geometry. In order to put him in a position to be able to wrestle with problems that are sure to crop up in estimating, I deem it expedient to arm him with rules and methods for obtaining areas, dimensions, and contents of all sorts of figures or solids he may meet with.

It is but just to say that these rules and methods can be found in many works, but it has been thought expedient to reproduce them here, so that the student may have them at hand when making use of this work for study or for practical estimating. The rules and problems are selected chiefly from educational works, and the tables have been prepared by competent authorities, and have been examined and corrected, where necessary, and made suitable to the work in hand.

It is presumed, at the outset, that the reader has some knowledge of arithmetic and is therefore able to follow without difficulty the problems that follow, which, after all, should offer no serious obstruction to a thorough knowledge of their qualities.

MENSURATION OF SUPERFICIES

Mensuration is that branch of mathematics by which we ascertain the contents or superficial areas, and the extension, solidities, and capacities of bodies.

The *area*, or superficial contents of any figure, is the measure of its surface, or the space contained within the bounds of that surface, without any regard to thickness.

In calculating the area, or the contents of any plane figure, some particular portion of surface is fixed upon as the *measuring unit*, with which the figure is to be compared.

This is commonly a *square*, the side of which is the unit of length, being an *inch*, or a *foot*, or a *yard*, or any other fixed quantity, according to the measure peculiar to different artists; and the area or contents of any figure is computed by the number of those squares contained in that figure.

For the same reason, determining the quantity of surface in a figure is called *squaring it;* that is, determining the square or number of squares to which it is equal.

In order to form correct estimates of the extent of surfaces and solids, various rules have been adopted, most of which, the most valuable and useful in practice, will be found accompanying their respective problems in the following treatise, and with which the mechanic may speedily perform all the calculations that ordinarily occur in the practical details of his business.

DEFINITIONS

The following definitions, which are similar in substance to those found in Euclid, are here inserted for the convenience of reference.

I. *Four-sided* figures are variously named, according to their relative position and length of their sides.

1. A *line* is length, without breadth or thickness.

2. *Parallel lines* are always at the same perpendicular distance and they never meet, though ever so far produced.

3. An *angle* is the inclination or opening of two lines, having different directions, and meeting in a point.

4. A *parallelogram* has its opposite sides parallel and equal.

5. A *rectangle*, or *right parallelogram*, has its opposite sides equal, and all its angles right angles.

6. A *square* is a figure whose sides are of equal length, and all its angles right angles.

7. A *rhomboid* has its opposite sides equal, and its angles oblique.

8. A *rhombus* is an equilateral rhomboid, having all its sides equal, but its angles oblique.

9. A *trapezoid* is a quadrilateral figure, having only two of its sides parallel.

10. A *trapezium* is an irregular figure, of four unequal sides and angles.

II. When figures have more than four sides, they are classed under the head of *Polygons*.

These again are either regular or irregular, according as their sides and angles are equal or unequal, and they are named from their number of sides or angles. Thus, a regular polygon has all its sides and angles equal.

A pentagon	has	five	sides
A hexagon	"	six	"
A heptagon	"	seven	"
An octagon	"	eight	"
A nonagon	"	nine	"
A decagon	"	ten	"
An undecagon	"	eleven	"
A dodecagon	"	twelve	"

III. A figure of three sides and angles is called a *triangle*, and receives particular denominations from the relations of its sides and angles.

1. An *equilateral triangle* is that whose three sides are equal.

2. The *height* of a triangle is the length of a perpendicular drawn from one of the angles to the opposite side.

3. An *isosceles triangle* is that which has only two sides equal.

4. The *height* of a four-sided figure is the perpendicular distance between two of its parallel sides.

OF FOUR-SIDED FIGURES

Problem I.—To find the area of a four-sided figure, whether it be a parallelogram, square, rhombus, or rhomboid.

Rule.—Multiply the length by the breadth or perpendicular height, and the product will be the area.

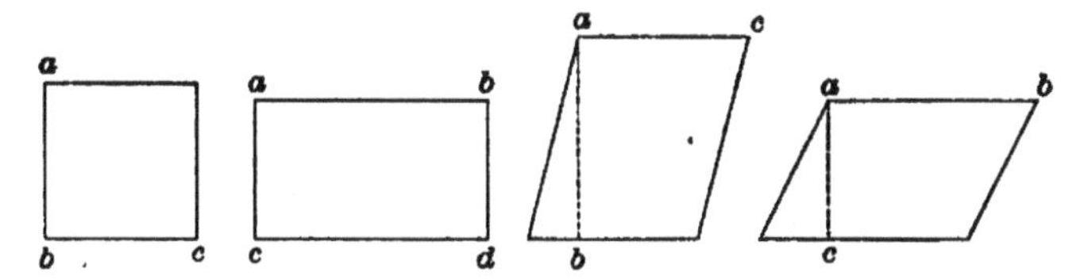

Example.—What is the area of a parallelogram, *a b c d*, whose length, *c d*, is 12 feet 3 inches, and whose breadth, *a c*, is 8 feet 6 inches?

BY DECIMALS.	BY DUODECIMALS.
Feet.	Feet.
12.25	12.3′
8.50	8.6′
61250	6. 1′ 6″
9800	98. 0′
104.1250 feet. Ans.	104. 1′ 6″. Ans.

NOTE. The fundamental problem, in the mensuration of superficies, is the very simple one of determining the area of a *right parallelogram*. The contents of other figures may readily be obtained by finding parallelograms which are equal to them.

Take any parallelogram, *a b c d*, and divide each of its sides, respectively, into as many equal parts as are expressed by the number of times they contain the linear measuring unit, and let all the opposite points of division be connected by right lines. Then it is evident that these lines divide the parallelogram into a number of squares, each equal to the superficial measuring unit, and that the number of these squares, or the area of the figure, is equal to the number of linear measuring units in the length, repeated as often as there are linear measuring units in the breadth or height; that is, equal to the length multiplied by the height, *which is the rule.*

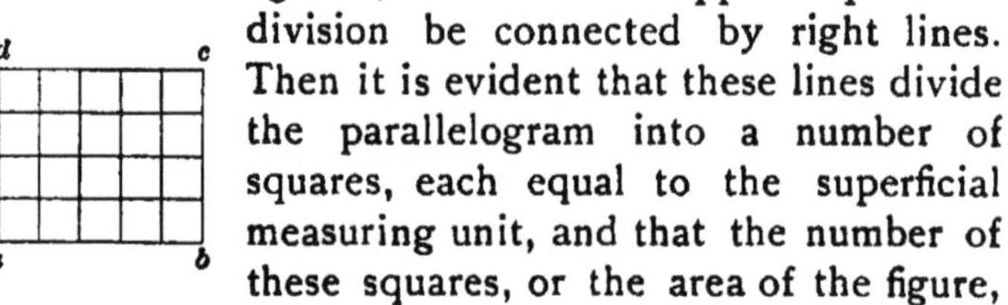

OF TRIANGLES

Problem II.—To find the area of a triangle.

Rule.—Multiply the length of one of the sides by the perpendicular falling upon it, and half the product will be the area. Or multiply half the side by the perpendicular.

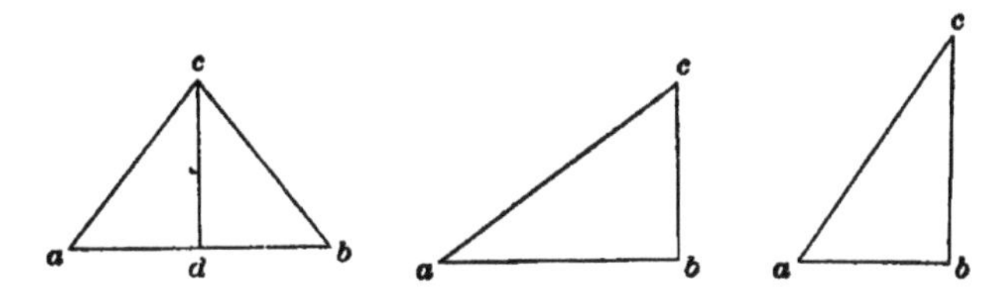

Example.—What is the area of a triangle whose base, *a b*, is 18 feet 4 inches, and height, *c d*, 11 feet 10 inches?

$$18.4 \times 11.10 \div 2 = 108 \text{ feet } 5\tfrac{2}{3} \text{ inches.}$$

Example 2.—How many square rods of land are there in a lot which is laid out in a right-angled triangle, the base measuring 19 rods, and the perpendicular breadth 15 rods? Ans. 142.5.

Case II.—To find the area of a triangle from the length of its sides.

Rule.—1. Add together the lengths of the three sides, and take half their sum.

2. From this half sum subtract each side separately.

3. Multiply together the half sum and each of the three remainders, and extract the square root of the product; the quotient will be the required area of the triangle.

Example.—If the sides of a triangle are 134,108 and 80 rods, what is the area?

134	161	161	161
108	134	108	80
80	27 1st rem.	53 2d rem.	81 3d rem.
322÷2=161 half sum.			

Then, to obtain the products, we have $161 \times 27 \times 53 \times 81 =$ 18661671: from which we find area$=\sqrt{18661671}=4319$ square rods.

To find the hypotenuse of a right-angled triangle, when the base and perpendicular are known.

1. Square each of the sides separately.
2. Add together these squares.
3. Extract the square root of the sum, which will be the hypotenuse.

Example.—The wall of a building, *b c*, on the bank of a river, *a b*, is 120 feet high, and the breadth of the river 210 feet: what is the length of a line, *a c*, which will reach from the top of the wall to the opposite bank of the river?

$$\overline{120}^2\times\overline{210}^2=58500 \text{ and } \sqrt{58500}=241.86 \text{ ft. Ans.}$$

To find one of the legs when the hypotenuse and the other leg are known.

Rule.—Subtract the square of the leg whose length is known, from the square of the hypotenuse, and the square root of their difference will be the answer.

Example.—The hypotenuse, *a c*, of a triangle is 53 yards, and the perpendicular, *b c*, 45 yards: what is the length of the base, *a b*?

$$\overline{53}^2-\overline{45}^2=784 \text{ and } \sqrt{784}=28 \text{ yds. Ans. } 28 \text{ yds.}$$

OF TRAPEZIUMS AND TRAPEZOIDS

Problem III.—To find the area of a trapezium.

Rule.—Divide the trapezium into triangles by drawing diagonals; and the sum of the areas of these triangles will be the area of the trapezium.

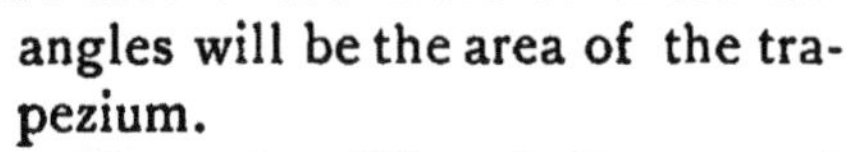

Example.—What is the area of a trapezium whose diagonal, *a c*, is 42 feet, and the two perpendiculars, *d e* and *b f*, 18 and 16 feet?

$$\left.\begin{array}{l}42\times 9=378\\ 42\times 8=336\end{array}\right\}=714 \text{ sq. ft. Ans.}$$

Problem IV.—To find the area of a trapezoid.

Rule.—Multiply the sum of the two parallel sides by the perpendicular distance between them, and half the product will be the area.

Example 1.—Required the area of the trapezoid, *a b c d*, having given *a b* = 321.51 feet, *d c* = 214.24 feet, and whose height is 171.16 feet.

We first find the sum of the sides, and then multiply it by the perpendicular height; after which, we divide the product by 2 for the area.

321.51+214.24=535.75=the sum of the parallel sides.
Then, 535.75×171.16=91698.97.
And, 91698.97÷2=45849.485. Ans.

OF REGULAR POLYGONS

Problem V.—To find the area of a regular polygon, or any regular figure.

Rule 1.—Multiply one of its sides into half its perpendicular distance from the center, and this product into the number of sides.

It is evident, on inspection, that a regular polygon contains as many equal triangles as the figure has sides.

Thus, the adjoining hexagon has six triangles, each equal to *a b c*. Now, the area of *a b c* is equal to the product of the side *a b* into ½ of *c d*. The area of the whole, therefore, is equal to this product multiplied into the *number* of sides.

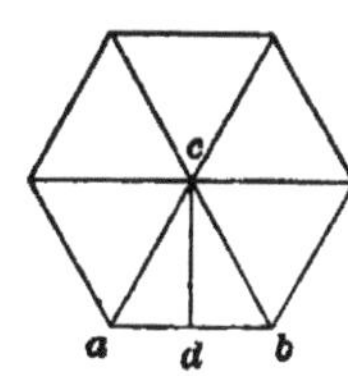

Example.—1. Required the area of a regular hexagon, each of whose sides, *a b*, etc., is 45 feet, and the perpendicular, *c d*, 24 feet.

We first multiply one side by ½ of the perpendicular, *c d*, and that product by the number of sides: this gives the area.

$48 \times 12 \times 6 = 3240$ ft. Ans.

To facilitate the measurement of polygons, the following table is constructed, showing the multipliers of the ten regular polygons, when the sides of each are equal to 1:

No. of sides.	Name of Polygon.	Angle.	Angle of Polygon.	Area of Multipliers	A	B	C
3	Triangle ..	120	60°	0.433012	2.	1.732	.5773
4	Square....	90	90	1.	1.41	1.414	.7071
5	Pentagon .	72	108	1.720477	1.238	1.175	.8506
6	Hexagon. .	60	120	2.598076	1.156	=Radius	Lgth of side
7	Heptagon .	51 3/7	128 4/7	3.633912	1.11	.8677	1.152
8	Octagon ..	45	135	4.828427	1.08	.7653	1.3065
9	Nonagon ..	40	140	6.181824	1.06	.6840	1.4619
10	Decagon ..	36	144	7.694208	1.05	.6180	1.6180
11	Undecagon	32 3/11	147 3/4	9.365640	1.04	.5634	1.7747
12	Dodecagon	30	150	11.196152	1.037	.5176	1.9318

Now, since the areas of similar polygons are to each other as the squares of their homologous sides, if the square of a side of a polygon be multiplied by the multiplier of the like figure, the product will be the area sought. And hence we have,

1^2 : tabular area : : any side squared : area.

To find the area of a regular polygon, when the side only is given.

Rule.—Multiply the square of the side by the multiplier opposite the name of the polygon in the above table, and the product will be the area.

Example.—What is the area of a regular decagon whose side is 87 feet?

$87^2 \times 7.694208 = 58237.46$. Ans.

ADDITIONAL USE OF THE ABOVE TABLE

The third and fourth columns of the table will greatly facilitate the construction of those figures with the aid of the sector. Thus, if it is required to describe an *octagon*, opposite to it, in the third column, is 45; then with the chord of 60 on the sector as radius, describe a circle, taking the length 45 on the same line of the sector; mark this distance off on the circumference, which, being repeated around the circle, will give the points of the side.

The fourth column gives the angle which any two adjoining sides of the respective figures make with each other.

Take the length of a perpendicular drawn from the center of one of the sides of a polygon, and multiply this by the numbers in column A; the product will be the radius of the circle that contains the figure.

The radius of a circle, multiplied by the number in column B, will give the length of the side of the corresponding figure which that circle will contain. The length of the side of a polygon, multiplied by the corresponding number in the column C, will give the radius of the circumscribing circle.

OF IRREGULAR BODIES

To find the area of an irregular polygon.

Rule.—Draw diagonals to divide the figure into trapeziums and triangles; find the area of each separately, and the sum of the whole will give the area required.

What is the area of the adjoining polygon, *a b c d e f g h*?

Let ac=20 rods.
" bp= 4 "
" ac=20 "
" hp= 6 "
" ce=25 "
" dp= 3 "
" fh=28 "
" gp= 7 "
" fh=28 "
" ep= 8 "
" hce=25 " each. 618.8 sq. rods. Ans.

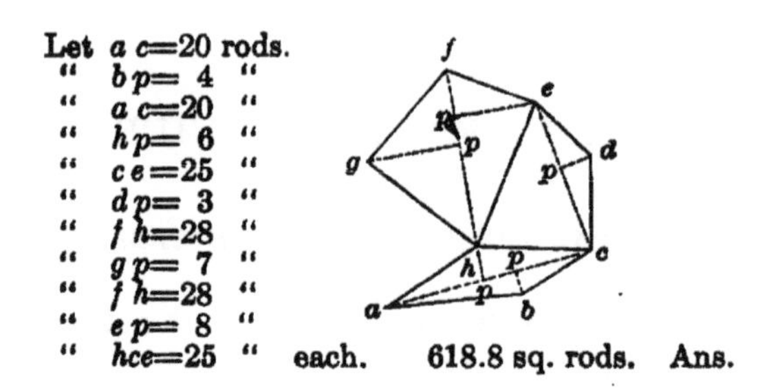

NOTE The triangle, hce, is solved by Problem II, Case II.

Problem VI.—To find the area of a long irregular figure, bounded on one side by a straight line.

Rule.—1. Measure the breadth in several places, and at equal distances from each other.

2. Add together all the different breadths, and half the sum of the two extremes.

3. Multiply this sum by the base line, and divide the product by the number of equal parts of the base.

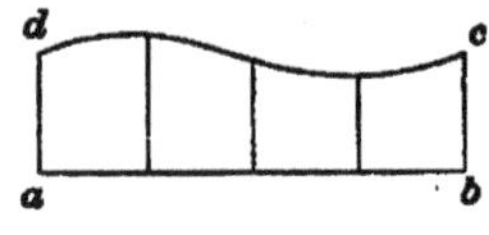

Example.—1. The breadths of an irregular figure, $a\ b\ c\ d$, at five equidistant places, being 8.2, 7.4, 9.2, 10.2, 8.6, and the whole length 39, required the area.

```
     8.2                            35.2=sum.
     8.6                             39
  2)16.8=sum of extremes.          3168
     8.4=mean of extremes.        1056
     7.4                        4)1372.8
     9.2                           343.2. Ans.
    10.2
    35.2 sum.
```

2. The length of an irregular figure being 84, and the breadths at six equidistant places, 17.4, 20.6, 14.2, 16.5, 20.1, 24.4, what is the area? 1550.64. Ans.

NOTE. If the perpendiculars or breadths be not at equal distances, add them together, and divide their sum by the number of them, for the mean breadth; then multiply the mean breadth by the length, and the product will be the whole area not far from the truth.

OF THE CIRCLE AND ITS PARTS

DEFINITIONS

1. A *circle* is a plane figure, bounded by a curved line, called the circumference, every part of which is equally distant from a certain point within, called the center.

2. A *diameter* of a circle is a straight line, passing through the center, and terminating at the circumference.

3. A *radius* or *semi-diameter* is a straight line, extending from the center to the circumference.

4. A *semi-circle* is one half of the circumference.

5. A *quadrant* is one quarter of the circumference.

6. An *arc* is any portion of the circumference.

7. A *chord* is a straight line, which joins the two extremes of an arc.

8. A *circular segment* is the space contained between an arc and its chord. The chord is sometimes called the *base* of the segment. The *height* of the segment is the perpendicular from the middle of the base to the arc.

9. A *circular sector* is the space contained between an arc and the two radii, drawn from the extremes of the arc.

10. A *circular zone* is the space contained between two parallel chords which form its bases.

11. A *circular ring* is the space between the circumferences of two concentric circles.

12. A *lune* or *crescent* is the space between two circular arcs, which intersect each other.

13. An *ellipse* or *oval* is a curve line, which returns into itself like a circle, but has two diameters of unequal length, the longest of which is called the transverse, and the shortest the conjugate axis.

Problem I.—To find the circumference of a circle when the diameter is given.

Rule.—Multiply the diameter by 3.1416, and the product will be the circumference. Or, multiply the diameter by 22, and divide the product by 7. Or, multiply the diameter by 355, and divide the product by 113.

NOTE.—The latter rule is a little more accurate than any other expressed in small numbers.

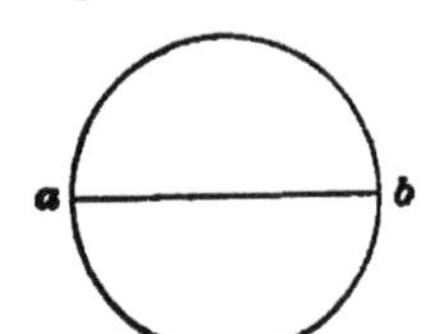

Example.—1. What is the circumference of a circle whose diameter, *a b*, is 40 feet?

$40 \times 3.1416 = 125.66$. Ans.

Example.—2. Required the circumference of a circle whose diameter is 73¾.

Ans. 231.6922.

NOTE.—See Table of Circumferences of Circles.

Problem II.—To find the diameter of a circle when the circumference is given.

Rule.—Divide the circumference by 3.1416, and the quotient will be the diameter. Or, multiply the circumference by 7, and divide the product by 22.

Example.—The circumference of a circle is 69.115 yards: what is the diameter?

$69.115 \div 3.1416 = 22$ yards.

The same result may be obtained more conveniently, by exchanging the *divisor*, 3.1416, for a *multiplier*, which will give the same answer, for, in the proportion 3.1416 : 1 :: Circ. : Diam., the fourth term may be directly found by dividing the second by the first, and multiplying the quotient into the third. Thus, $1 \div 3.1416 = 0.31831$. Therefore, if the circumference of any circle be *multiplied* by the decimal .31831, the product will be the diameter.

In many cases there will be a decided saving of labor by exchanging the *divisor* for a *multiplier*, as will be seen in the following example:

Example.—What is the diameter of a circle whose circumference is 50?

$$50 \times .31831 = 15.91550.$$

NOTE.—As multiplication is more easily performed than division, this last method is decidedly the more preferable.

Problem III.—To find the area of a circle when the diameter and circumference are both known.

Rule.—Multiply the square of the diameter by .7854. Or, the square of the circumference by .07958. Or, multiply the circumference by the diameter, and divide the product by 4; in either case the product will be the area.

Example.—1. Required the number of square inches in a piston whose diameter is 12½ inches.

$\overline{12\tfrac{1}{2}}^2 = 12.5 \times 12.5 = 156.25$, and $156.25 \times .7854 = 122.71$ sq. in. Ans.

2. The piston of the railroad engine Boston is 15 inches diameter: how many square inches does it contain? 176.71. Ans.

NOTE.—The reason of this rule will appear by considering that if the circumference of a circle be 1, the diameter will=0.31831 (Prob. II), and ¼ of this diameter into the circumference is 0.7958 =area. (See Table of Areas of Circles.)

Problem IV.—I. To find the length of an arc of a

circle, when either the number of degrees which it contains, or the radius, chord, and height are given.

Rule.—Multiply the number of degrees in the arc by the decimal .01745, and that product by the radius of the circle. Or, from 8 times the chord of half the arc, subtract the chord of the whole arc, and $\frac{1}{3}$ of the remainder will be the length of the arc, nearly. Or, as 3 is to the number of degrees in the arc, so is .05236 times the radius to its length.

Example.—1. What is the length of an arc of 40 degrees, in a circle whose radius, *a c*, is 12 feet?

$$.0745 \times 40 \times 12 = 8.376 = \text{length of the arc.}$$

2. What is the length of an arc whose chord, *a b*, is 120, and whose height, *p d*, is 45?

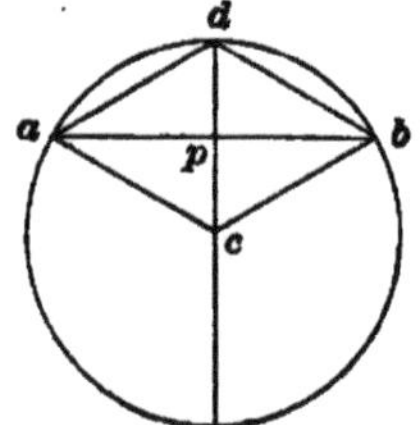

$120 \div 2 = 60 = \frac{1}{2}$ chord of the arc.
And $\overline{60}^2 = 3600$
" $\overline{45}^2 = 2025$
$\overline{5625}$ = sum of the squares.
Then $\sqrt{5625} = 75$ = chord of $\frac{1}{2}$ the arc.
And $75 \times 8 - 120 \div 3 = 160$. Ans.

NOTE.—The chord of half the arc is equal to the square root of the sum of the squares of the height and half the chord of the whole arc.

II. When the chord of the arc and the chord of half the arc are given.

Rule.—From the square of the chord of half the arc subtract the square of half the chord of the entire arc; the remainder will be the square of the versed sine. Then proceed as before.

NOTE.—The square root of the sum of the squares of the versed sine or height, and half the chord of the entire arc is equal to the chord of half the arc.

III. When the diameter and the versed sine of half the arc are given.

Rule.—From 60 times the diameter subtract 27 times the versed sine, and *reserve* the number. Multiply the diameter by the versed sine, and the square root of the product will be the *chord* of half the arc. Multiply twice the chord of half the arc by 10 times the versed sine, divide the product by the *reserved number*, and add the quotient to twice the chord of half the arc; the sum will be the length of the arc, very nearly.

TABLE OF THE RELATIVE PROPORTIONS OF THE CIRCLE, ITS EQUAL AND INSCRIBED SQUARES

1. The	diameter of a circle	× .8862	=side of an equal square.
2. "	circumference "	× .2821	
3. "	diameter "	× .7071	=side of an inscribed sq.
4. "	circumference "	× .2251	
5. "	arc "	× .6366	=contents of inscribed sq.
6. "	side of inscribed square	× 1.4142	=diam. circumscrib'g cir.
7. "	side of inscribed square	× 4.443	=circum. circumscrib'g cir.
8. "	side of a square	× 1.128	=diam. of an equal circle.
9. "	side of a square	× 3.545	=circum. of an equal sq.

Problem V.—To find the side of a square inscribed in a circle, from its circumference or diameter.

Rule.—Multiply the diameter by .7071=the side of the inscribed square. Or, multiply the circumference by .2251=side of the inscribed square.

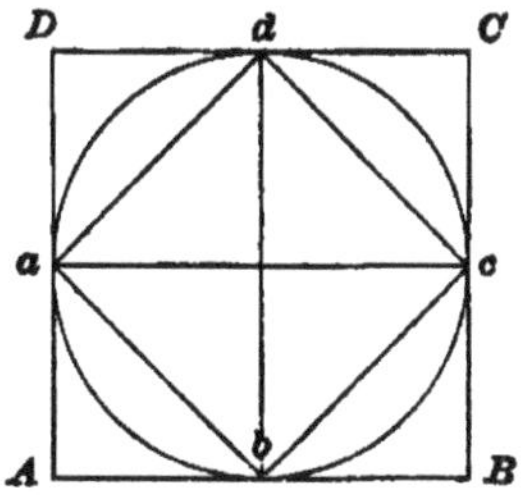

Example.—1. The circumference of a circle is 68 inches: what is the side of the inscribed square?

68×.2251=15.30 inches. Ans.

2. The diameter of a tree is 37½ inches at the small end: what is the measure of the side of the greatest square which can be sawed from it?

37.5×.7071=26.51 inches Ans.

NOTE.—The *area* of a circle is to the area of the *circumscribed square* as .7854 is to 1, and to that of the *inscribed square* as .7854 is to ½. If the reader will examine the above figure, he will see that the square, *A B C D*, which is circumscribed about the circle, is equal to the square of the diameter of the circle, since the diameter, *a c*, equals the side *A B*, and *A B* squared gives the area of the square *A B C D;* also, that the inscribed square, *abcd*, is just ½ of the circumscribed square. Since each of the triangles into which the inscribed square is divided is precisely half of each of the four squares into which the circumscribed square, *A B C D*, is divided. That is, the inscribed square contains only 4 right-angled triangles, while the circumscribed square contains 8. Consequently, the square described within a circle is precisely half of the square described without it.

Problem VI.—To find the area of a sector of a circle.

Rule.—1. Find the length of the arc by problem vii.

2. Multiply the length of the arc thus found, by half the length of the radius, and the product will be the area.

Or, as 360 degrees is to the number of degrees in the arc of the sector, so is the area of the circle to the area of the sector.

NOTE.—If the diameter of radius is not given, add the square of half the chord of the arc to the square of the versed sine of half the arc, and divide the sum by the versed sine; the quotient will be the diameter.

It is manifest that the area of the sector has the same ratio to the area of the circle which the number of *degrees* in the arc has to the number of degrees in the whole circumference; and the rule for finding the area of the sector, is the same as that for finding the area of the whole circle.

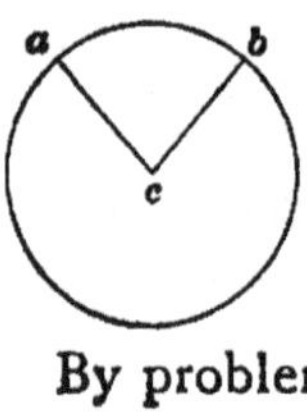

a b c

Example.—What is the area of a sector of a circle, *a c b*, in which the radius, *a c*, is 25 and the arc of 26 degrees?

By problem vii. Rule 3.

As, 3 : 26 : : 25×.05236 : 11.344; and 11.344×12½=141.8. Ans.

Problem VII.—To find the area of the segment of a circle.

Rule.—1. To the chord of the whole arc add ⅓ of the chord of half the arc.

2. Then multiply the sum by the versed sine, or height of the segment, and $\frac{4}{10}$ of the product will be the area of the segment, very nearly.

3. Divide the height or versed sine by the diameter of the circle, and find the quotient in the column of versed sines. (See table.) Then take out the corresponding area in the next column on the right hand, and multiply it by the square of the diameter for the answer.

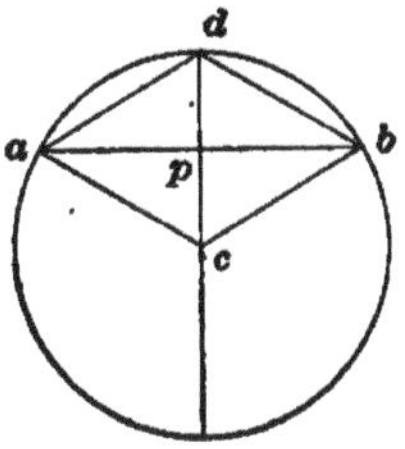

Example.—1. Required the area of a circular segment whose chord, $a\,b$, = 24, and whose radius, $c\,a$, = 20 feet?

$\overline{ca}^2 - \overline{ap}^2 = \overline{cp}^2 = \sqrt{400-144} = 16 = cp.$

$cd - cp = dp = 20 - 16 = 4 =$ height of segment.

$\overline{ap}^2 + \overline{pd}^2 = \overline{ad}^2 = \sqrt{144+16} = 12.64911 =$ chord ad.

24 =the chord of the segment.

12.64911=chord of ½ the segment.

4.21637=⅓ of the chord of ½ the arc.

40.86548=the height of the segment.

163.46192×4÷10=65.384768=area of the segment. Ans.

(See Table of Areas of the Segments of Circles.)

OF LUNES

Problem VIII.—To find the area of a lune or crescent.

Rule.—Find the difference of the two segments which are between the arcs of the crescent and its chord for the area.

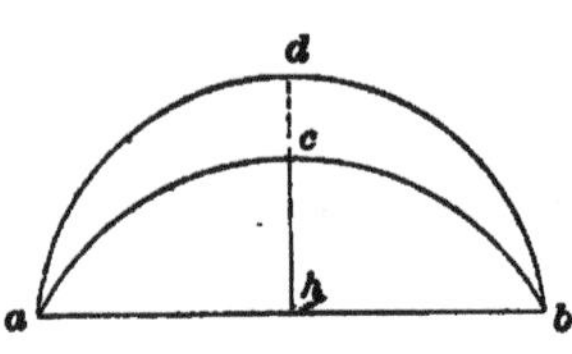

Example.—The chord of two segments, $a\ b$, is 72, and the height of the greater segment, $h\ d$, is 30, and of the lesser, $h\ c$, 20: what is the area of the crescent?

$\overline{30^2}+\overline{36^2}=2196$ and $\sqrt{2196}=46.8$=chord of half the arc.

And $46.8\times\frac{4}{3}=62.4$: Then, $62.4+72\times30\times\frac{4}{10}=1612.8$=area of segment, *abd*.

Again, $\overline{20^2}+\overline{36^2}=1696$ and $\sqrt{1696}=41.2$=chord of ½ arc.

Then, $41.2\frac{4}{3}=50.8$, and $50.8+72\times20\times\frac{4}{10}=982.4$=area of segment, *a b c*.

The difference of these areas is 630.4=the area of the lune or crescent.

NOTE.—If upon the three sides of a right-angled triangle, as diameters, semicircles be described, two lunes will be formed, whose united areas will be equal to the area of the triangle.

Problem IX.—To find the area of a circular zone.

Rule.—From the area of the whole circle, subtract the areas of the two segments on the sides of the zone.

If from the whole circle there be taken the two segments, *a b c* and *d f g*, there will remain the circular zone, *a c f d*.

Example.—1. What is the area of the zone, *a c f d*, if *a c* is 7.75, *d f* 6.93, and the diameter of the circle 8?

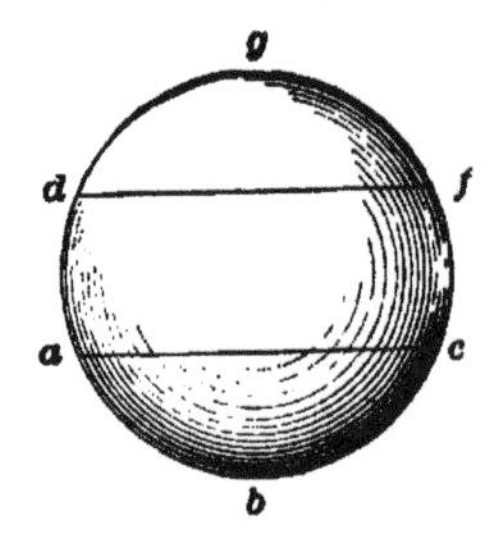

50.26=area of the whole circle.
17.23=area of the segment, *a b c*.
9.82=area of the segment, *d f g*.

27.05

And 50.26—27.05=23.21=area of the zone, *a c f d*.

Problem X.—To find the area of a ring included between the circumferences of two concentric circles.

Rule.—1. Square the diameter of each circle, and subtract the square of the less from that of the greater.

2. Multiply the difference of the squares by the decimal .7854, and the product will be the area.

Or, multiply the product of the *sum* and *difference* of the two diameters by .7854.

Example.—If the diameter of the outer circle, *a b*, be 221, and the inner circle, *d c*, 106, what is the area of the ring?

First, $\overline{221}^2 \times .7854 =$ 38359.72
And, $\overline{106}^2 \times .7854 =$ 8824.75
Ans. 29534.97

NOTE.—The area of each of these circles is equal to the square of the diameter multiplied by .7854 (Prob. 3). And the difference of these squares is equal to the product of the *sum* and *difference* of the diameters. Therefore, the area of the ring is equal to the product of the sum and difference of the two diameters, multiplied by .7854.

OF ELLIPSES

Problem XI.—To find the area of an ellipse.

Rule.—Multiply the longer axis by the shorter, and the product, multiplied by the decimal .7854, will be the area required.

NOTE.—A common and more scientific name for the longer axis of an ellipse, is the *transverse* or *major*, and for the shorter, the *conjugate* or *minor*.

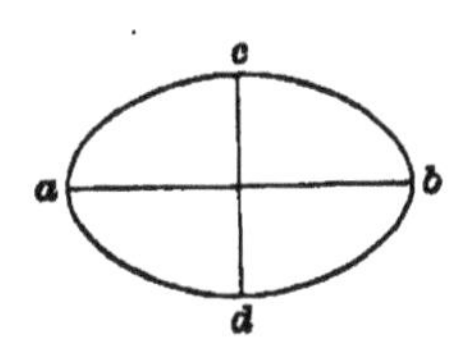

Example.—1. What is the area of an ellipse whose longer axis, *a b*, is 70 feet, and whose shorter, *d c*, is 50 feet?

$ab \times dc = 70 \times 50 = 3500.$
Then, $3500 \times .7854 = 2748.9 =$ area.

2. What is the area of an ellipse whose axes are 16 and 12? 150.79.. Ans.

Problem XII.—To find the circumference of an ellipse.

Rule.—Square the two axes, and multiply the square root of half their sum by 3.14159; the product will be the circumference, nearly.

Example.—What is the circumference of an ellipse whose transverse and conjugate axes are 16 and 18 feet?

$\overline{16}^2+\overline{18}^2=580$=sum of the squares of the axes.
And, 290=half sum.
Then, $\sqrt{290}\times3.14159=53.498$=circumference.

Problem XIII.—To find the area of an elliptic segment, cut off by a line perpendicular to either axis.

Rule.—Find the area of a corresponding circular segment, having the same height and the same vertical axis or diameter. Then say, as the vertical axis is to the other axis, parallel to the segment's base, so is the area of the circular segment before found, to the area of the elliptic segment sought.

Example.—The height of an elliptic segment is 10, and the axes 25 and 35 respectively: what is the area?

$10\div35=.2857$ tabular versed sine and segment=.18452.
And, $.18452\times35^2=249.98$.
Then, 25 : 35 : : 249.98 : 349.97=area.

Problem XIV.—To find the area of a parabola.

Rule.—Multiply the base by the height, and two-thirds of the product will be the area.

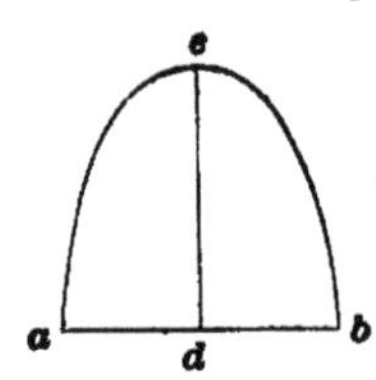

Example.—What is the area of a parabola, whose base, *a b*, is 26 inches, and height, *d e*, 18 inches?

$26\times18=468$=product of base and height.
$468\times\frac{2}{3}=312$=area in square inches.
Then $312\div144=2\frac{1}{6}$ square feet. Ans.

Problem XV.—To find the area of a frustum of a parabola, cut off by a line drawn parallel to the base.

Rule.—Multiply the difference of the cubes of the two ends of the frustum by twice its altitude, and divide the product by three times the difference of their squares.

Example.—What is the area of a frustum of a parabola whose height, *c b*, is 12 feet, and its upper end, *a e*, 12 feet, and its base, *d f*, 20 feet?

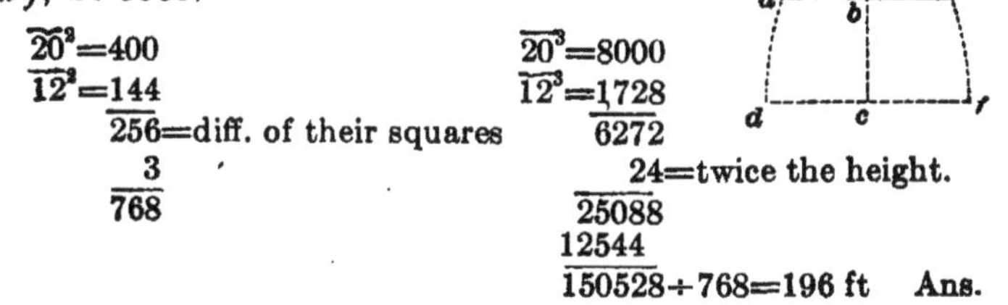

$\overline{20}^2=400$ $\qquad$ $\overline{20}^3=8000$

$\overline{12}^2=144$ $\qquad$ $\overline{12}^3=1728$

256=diff. of their squares $\qquad$ 6272

3 $\qquad$ 24=twice the height.

768 $\qquad$ 25088

12544

150528÷768=196 ft **Ans.**

OF HYPERBOLAS

Problem XVI.—To find the area of a hyperbola.

Rule.—To five-sevenths of the abscissa, *v e*, add the transverse diameter; multiply the sum by the abscissa, and extract the square root of the product. Then, multiply the transverse diameter, *v g*, by the abscissa, *v e*, and extract the square root of that product. Then, to 21 times the first root, add 4 times the second root; multiply the sum by double the product of the conjugate and abscissa, and divide by 75 times the transverse; this will give the area, nearly.

Example.—What is the area of a Hyperbola, *d f v*, whose transverse diameter, *v g*, is 80, and conjugate, *d f*, 50, and whose abscissa, *v e*, is 45?

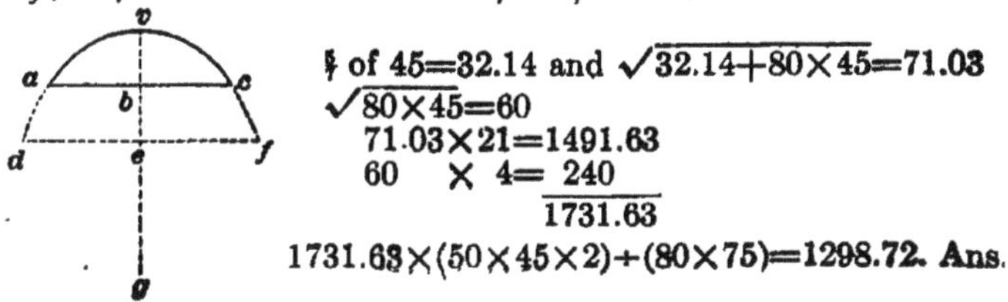

$\frac{5}{7}$ of 45=32.14 and $\sqrt{32.14+80\times45}=71.03$

$\sqrt{80\times45}=60$

$71.03\times21=1491.63$

$60 \times 4= 240$

1731.63

$1731.63\times(50\times45\times2)\div(80\times75)=1298.72$. **Ans.**

TABLE OF THE AREAS OF THE SEGMENTS OF A CIRCLE,

WHOSE DIAMETER IS UNITY AND SUPPOSED TO BE DIVIDED INTO 1000 EQUAL PARTS

V'rs'd Sine	Area of Segment	V'rs'd Sine	Area of Segment	V'rs'd Sine	Area of Segment	V'rs'd Sine	Area of Segment	V'rs'd Sine	Area of Segment
.001	.00004	.039	.01014	.077	.02782	.115	.05016	.153	.07602
.002	.00011	.040	.01053	.078	.02835	.116	.05080	.154	.07674
.003	.00021	.041	.01093	.079	.02889	.117	.05144	.155	.07746
.004	.00033	.042	.01133	.080	.02943	.118	.05209	.156	.07819
.005	.00047	.043	.01173	.081	.02997	.119	.05273	.157	.07892
.006	.00061	.044	.01214	.082	.03052	.120	.05338	.158	.07964
.007	.00077	.045	.01255	.083	.03107	.121	.05403	.159	.08038
.008	.00095	.046	.01297	.084	.03162	.122	.05468	.160	.08111
.009	.00113	.047	.01339	.085	.03218	.123	.05534	.161	.08184
.010	.00132	.048	.01381	.086	.03274	.124	.05600	.162	.08258
.011	.00153	.049	.01424	.087	.03330	.125	.05666	.163	.08332
.012	.00174	.050	.01468	.088	.03387	.126	.05732	.164	.08405
.013	.00196	.051	.01511	.089	.03444	.127	.05799	.165	.08480
.014	.00219	.052	.01556	.090	.03501	.128	.05865	.166	.08554
.015	.00243	.053	.01600	.091	.03558	.129	.05932	.167	.08628
.016	.00268	.054	.01645	.092	.03616	.130	.05999	.168	.08703
.017	.00294	.055	.01691	.093	.03674	.131	.06067	.169	.08778
.018	.00320	.056	.01736	.094	.03732	.132	.06134	.170	.08853
.019	.00347	.057	.01783	.095	.03790	.133	.06202	.171	.08928
.020	.00374	.058	.01829	.096	.03849	.134	.06270	.172	.09004
.021	.00403	.059	.01876	.097	.03908	.135	.06338	.173	.09079
.022	.00432	.060	.01923	.098	.03968	.136	.06407	.174	.09155
.023	.00461	.061	.01971	.099	.04027	.137	.06476	.175	.09231
.024	.00492	.062	.02019	.100	.04087	.138	.06544	.176	.09307
.025	.00523	.063	.02068	.101	.04147	.139	.06614	.177	.09383
.026	.00554	.064	.02116	.102	.04208	.140	.06683	.178	.09460
.027	.00586	.065	.02165	.103	.04268	.141	.06752	.179	.09536
.028	.00619	.066	.02215	.104	.04329	.142	.06822	.180	.09613
.029	.00652	.067	.02265	.105	.04390	.143	.06892	.181	.09690
.030	.00686	.068	.02315	.106	.04452	.144	.06962	.182	.09767
.031	.00720	.069	.02365	.107	.04513	.145	.07032	.183	.09844
.032	.00755	.070	.02416	.108	.04575	.146	.07103	.184	.09922
.033	.00791	.071	.02468	.109	.04638	.147	.07174	.185	.09999
.034	.00827	.072	.02519	.110	.04700	.148	.07245	.186	.10077
.035	.00863	.073	.02571	.111	.04763	.149	.07316	187	.10155
.036	.00900	.074	.02623	.112	.04826	.150	.07387	.188	.10233
.037	.00938	.075	.02676	.113	.04889	.151	.07458	.189	.10311
.038	.00976	.076	.02728	.114	.04952	.152	.07530	190	.10390

V'rs'd Sine	Area of Segment	V'rs'd Sine	Area of Segment	V'rs'd Sine	Area of Segment	V'rs'd Sine	Area of Segment	V'rs'd Sine	Area of Segment
.191	.10468	.240	.14494	.289	.18814	.338	.23358	.387	.28066
.192	.10547	.241	.14579	.290	.18904	.339	.23452	.388	.28164
.193	.10626	.242	.14665	.291	.18995	.340	.23547	.389	.28261
.194	.10705	.243	.14751	.292	.19086	.341	.23642	.390	.28359
.195	.10784	.244	.14837	.293	.19177	.342	.23736	.391	.28456
.196	.10863	.245	.14923	.294	.19268	.343	.23831	.392	.28554
.197	.10943	.246	.15009	.295	.19359	.344	.23926	.393	.28652
.198	.11022	.247	.15095	.296	.19450	.345	.24021	.394	.28749
.199	.11102	.248	.15181	.297	.19542	.346	.24116	.395	.28847
.200	.11182	.249	.15268	.298	.19633	.347	.24212	.396	.28945
.201	.11262	.250	.15354	.299	.19725	.348	.24307	.397	.29043
.202	.11342	.251	.15441	.300	.19816	.349	.24402	.398	.29141
.203	.11423	.252	.15528	.301	.19908	.350	.24498	.399	.29239
.204	.11503	.253	.15614	.302	.20000	.351	.24593	.400	.29336
.205	.11584	.254	.15701	.303	.20092	.352	.24688	.401	.29434
.206	.11665	.255	.15789	.304	.20184	.253	.24784	.402	.29533
.207	.11746	.256	.15876	.305	.20276	.354	.24880	.403	.29631
.208	.11827	.257	.15963	.306	.20368	.355	.24975	.404	.29729
.209	.11908	.258	.16051	.307	.20460	.356	.25071	.405	.29827
.210	.11989	.259	.16138	.308	.20552	.357	.25167	.406	.29925
.211	.12071	260	.16226	.309	.20645	.358	.25263	.407	.30023
.212	.12152	.261	.16314	310	.20737	.359	.25359	.408	.30122
.213	.12234	.262	.16401	.311	.20830	.360	.25455	.409	.30220
.214	.12316	.263	.16489	.312	.20922	.361	.25551	.410	.30318
.215	.12398	.264	.16578	.313	.21015	.362	.25647	.411	.30417
.216	.12481	.265	.16666	.314	.21108	.363	.25743	.412	.30515
.217	.12563	.266	.16754	.315	.21201	.364	.25839	.413	.30614
.218	.12645	.267	.16843	.316	.21294	.365	.25935	.414	.30712
.219	.12728	.268	.16931	.317	.21387	.366	.26032	.415	.30811
.220	.12811	.269	.17020	.318	.21480	.367	.26128	.416	.30909
.221	.12894	.270	.17108	.319	.21573	.368	.26224	.417	.31008
.222	.12977	.271	.17197	.320	.21666	.369	.26321	.418	.31106
.223	.13060	.272	.17286	.321	.21759	.370	.26417	.419	.31205
.224	.13143	.273	.17375	.322	.21853	.371	.26514	.420	.31304
.225	.13227	.274	.17464	.323	.21946	.372	.26611	.421	.31402
.226	.13310	.275	.17554	.324	.22040	.373	.26707	.422	.31501
.227	.13394	.276	.17643	.325	.22134	.374	.26804	.423	.31600
.228	.13478	.277	.17733	.326	.22227	.375	.26901	.424	.31699
.229	.13562	.278	.17822	.327	.22321	.376	.26998	.425	.31798
.230	.13646	.279	.17912	.328	.22415	.377	.27095	.426	.31897
.231	.13730	.280	.18001	.329	.22509	.378	.27192	.427	.31995
.232	.13815	.281	.18091	.330	.22603	.379	.27289	.428	.32094
.233	.13899	.282	.18181	.331	.22697	.380	.27386	.429	.32193
.334	.13984	.283	.18271	.332	.22791	.381	.27483	.430	.32292
.235	.14068	.284	.18361	.333	.22885	.382	.27580	.431	.32391
.236	.14153	.285	.18452	.334	.22980	.383	.27677	.432	.32490
.237	.14238	.286	.18542	.335	.23074	.384	.27774	.433	.32590
.238	.14323	.287	.18632	.336	.23168	.385	.27872	.434	.32689
.239	.14409	288	.18723	.337	.23263	.386	.27969	.435	.32788

V'rs'd Sine	Area of Segment	V'rs'd Sine	Area of Segment	V'rs'd Sine	Area of Segment	V'rs'd Sine	Area of Segment	V'rs'd Sine	Area of Segment
.436	.32887	.449	.34178	.462	.35473	.475	.36770	.488	.38070
.437	.32986	.450	.34278	.463	.35573	.476	.36870	.489	.38169
.438	.33085	.451	.34377	.464	.35673	.477	.36970	.490	.38269
.439	.33185	.452	.34477	.465	.35772	.478	.37070	.491	.38369
.440	.33284	.453	.34576	.466	.35872	.479	.37170	.492	.38469
.441	.33383	.454	.34676	.467	.35972	.480	.37270	.493	.38569
.442	.33482	.455	.34775	.468	.36072	.481	.37370	.494	.38669
.443	.33582	.456	.34875	.469	.36171	.482	.37470	.495	.38769
.444	.33681	.457	.34975	.470	.36271	.483	.37570	.496	.38869
.445	.33781	.458	.35074	.471	.36371	.484	.37670	.497	.38969
.446	.33880	.459	.35174	.472	.36471	.485	.37770	.498	.39069
.447	.33979	.460	.35274	.473	.36571	.486	.37870	.499	.39169
.448	.34079	.461	.35373	.474	.36671	.487	.37970	.500	.39269

USE OF THE ABOVE TABLE

To find the area of a segment of a circle.

Rule.—Divide the height, or versed sine, by the diameter of the circle, and find the quotient in the column of versed sines.

Then take out the corresponding area, in the next column on the right hand, and multiply it by the square of the diameter; this will give the area of the segment.

Example.—Required the area of a segment of a circle, whose height is $3\frac{1}{4}$ feet, and the diameter of the circle 50 feet?

$$3\tfrac{1}{4}=3.25;\ \text{and}\ 3.25\div 50=.065.$$

.065, as per table=.021659; and $.021659\times 50^2=54.147500$, the area required.

Approximating rule to find the area of a segment of a circle.

Rule.—Multiply the chord of the segment by the versed sine, divide the product by 3, and multiply the remainder by 2.

Cube the height, or versed sine, find how often twice the length of the chord is contained in it, and add the quotient to the former product; this will give the area of the segment, very nearly.

Example.—Required the area of the segment of a circle, the chord being 12, and the versed sine 2.

$$12\times 2=24;\ 24\div 3=8;\ \text{and}\ 8\times 2=16.$$

$$2^3\div 24=.3333.$$

Hence $16+.3333=16.3333$, the area of the segment, very nearly.

TABLE OF THE AREAS OF THE ZONES OF A CIRCLE

V'rs'd Sine	Area of Segment	V'rs'd Sine	Area of Segment	V'rs'd Sine	Area of Segment	V'rs'd Sine	Area of Segment	V'rs'd Sine	Area of Segment
.001	.00100	.044	.04394	.087	.08655	.130	.12852	.173	.16948
.002	.00200	.045	.04494	.088	.08754	.131	.12948	.174	.17042
.003	.00300	.046	.04593	.089	.08852	.132	.13045	.175	.17135
.004	.00400	.047	.04693	.090	.08951	.133	.13141	.176	.17229
.005	.00500	.048	.04792	.091	.09049	.134	.13237	.177	.17323
.006	.00600	.049	.04892	.092	.09147	.135	.13334	.178	.17416
.007	.00700	.050	.04991	.093	.09246	.136	.13430	.179	.17510
.008	.00800	.051	.05091	.094	.09344	.137	.13526	.180	.17603
.009	.00900	.052	.05190	.095	.09442	.138	.13622	.181	.17696
.010	.01000	.053	.05290	.096	.09540	.139	.13718	.182	.17789
.011	.01100	.054	.05389	.097	.09638	.140	.13814	.183	.17882
.012	.01199	.055	.05489	.098	.09736	.141	.13910	.184	.17975
.013	.01299	.056	.05588	.099	.09835	.142	.14006	.185	.18068
.014	.01399	.057	.05687	.100	.09933	.143	.14102	.186	.18161
.015	.01499	.058	.05787	.101	.10030	.144	.14198	.187	.18254
.016	.01599	.059	.05886	.102	.10128	.145	.14294	.188	.18347
.017	.01699	.060	.05985	.103	.10226	.146	.14389	.189	.18439
.018	.01799	.061	.06084	.104	.10324	.147	.14485	.190	.18532
.019	.01899	.062	.06184	.105	.10422	.148	.14581	.191	.18624
.020	.01999	.063	.06283	.106	.10520	.149	.14676	.192	.18717
.021	.02099	.064	.06382	.107	.10617	.150	.14771	.193	.18809
.022	.02199	.065	.06481	.108	.10715	.151	.14867	.194	.18901
.023	.02299	.066	.06580	.109	.10813	.152	.14962	.195	.18993
.024	.02399	.067	.06679	.110	.10910	.153	.15057	.196	.19085
.025	.02499	.068	.06779	.111	.11008	.154	.15153	.197	.19177
.026	.02598	.069	.06878	.112	.11105	.155	.15248	.198	.19269
.027	.02698	.070	.06977	.113	.11203	.156	.15343	.199	.19361
.028	.02798	.071	.07076	.114	.11300	.157	.15438	.200	.19453
.029	.02898	.072	.07175	.115	.11397	.158	.15533	.201	.19544
.030	.02998	.073	.07274	.116	.11495	.159	.15627	.202	.19636
.031	.03098	.074	.07372	.117	.11592	.160	.15722	.203	.19727
.032	.03197	.075	.07471	.118	.11689	.161	.15817	.204	.19819
.033	.03297	.076	.07570	.119	.11786	.162	.15911	.205	.19910
.034	.03397	.077	.07669	.120	.11883	.163	.16006	.206	.20001
.035	.03497	.078	.07768	.121	.11980	.164	.16101	.207	.20092
.036	.03596	.079	.07867	.122	.12077	.165	.16195	.208	.20183
.037	.03696	.080	.07965	.123	.12174	.166	.16289	.209	.20274
.038	.03796	.081	.08064	.124	.12271	.167	.16384	.210	.20365
.039	.03896	.082	.08163	.125	.12368	.168	.16478	.211	.20455
.040	.03995	.083	.08261	.126	.12465	.169	.16572	.212	.20546
.041	.04095	.084	.08360	.127	.12562	.170	.16666	.213	.20637
.042	.04195	.085	.08458	.128	.12658	.171	.16760	.214	.20727
.043	.04294	.086	.08557	.129	.12755	.172	.16854	.215	.20817

V'rs'd Sine	Area of Segment	V'rs'd Sine	Area of Segment	V'rs'd Sine	Area of Segment	V'rs'd Sine	Area of Segment	V'rs'd Sine	Area of Segment
.216	.20908	.265	.25201	.314	.29192	.363	.32793	.412	.35882
.217	.20998	.266	.25285	.315	.29270	.364	.32862	.413	.35939
.218	.21088	.267	.25370	.316	.29347	.365	.32931	.414	.35995
.219	.21178	.268	.25454	.317	.29425	.366	.32999	.415	.36051
.220	.21268	.269	.25539	.318	.29502	.367	.33067	.416	.36107
.221	.21357	.270	.25623	.319	.29579	.368	.33135	.417	.36162
.222	.21447	.271	.25707	.320	.29656	.369	.33202	.418	.36217
.223	.21536	.272	.25791	.321	.29733	.370	.33270	.419	.36272
.224	.21626	.273	.25875	.322	.29809	.371	.33337	.420	.36326
.225	.21715	.274	.25959	.323	.29886	.372	.33404	.421	.36380
.226	.21805	.275	.26042	.324	.29962	.373	.33470	.422	.36434
.227	.21894	.276	.26126	.325	.30038	.374	.33537	.423	.36487
.228	.21983	.277	.26209	.326	.30114	.375	.33603	.424	.36541
.229	.22072	.278	.26292	.327	.30190	.376	.33669	.425	.36593
.230	.22161	.279	.26375	.328	.30265	.377	.33735	.426	.36646
.231	.22249	.280	.26458	.329	.30341	.378	.33801	.427	.36698
.232	.22335	.281	.26541	.330	.30416	.379	.33866	.428	.36750
.233	.22426	.282	.26624	.331	.30491	.380	.33931	.429	.36801
.234	.22515	.283	.26706	.332	.30566	.381	.33996	.430	.36853
.235	.22603	.284	.26788	.333	.30641	.382	.34060	.431	.36904
.236	.22691	.285	.26871	.334	.30715	.383	.34125	.432	.36954
.237	.22780	.286	.26953	.335	.30789	.384	.34189	.433	.37004
.238	.22868	.287	.27035	.336	.30864	.385	.34253	.434	.37054
.239	.22955	.288	.27171	.337	.30937	.386	.34317	.435	.37104
.240	.23043	.289	.27198	.338	.31011	.387	.34380	.436	.37153
.241	.23131	.290	.27280	.339	.31085	.388	.34443	.437	.37201
.242	.23218	.291	.27361	.340	.31158	.389	.34506	.438	.37250
.243	.23306	.292	.27442	.341	.31231	.390	.34569	.439	.37298
.244	.23393	.293	.27523	.342	.31305	.391	.34631	.440	.37346
.245	.23480	.294	.27604	.343	.31377	.392	.34694	.441	.37393
.246	.23568	.295	.27685	.344	.31450	.393	.34756	.442	.37440
.247	.23655	.296	.27766	.345	.31523	.394	.34817	.443	.37486
.248	.23741	.297	.27846	.346	.31595	.395	.34879	.444	.37533
.249	.23828	.298	.27927	.347	.31667	.396	.34940	.445	.37578
.250	.23915	.299	.28007	.348	.31739	.397	.35001	.446	.37624
.251	.24001	.300	.28087	.349	.31811	.398	.35061	.447	.37669
.252	.24088	.301	.28167	.350	.31882	.399	.35122	.448	.37713
.253	.24174	.302	.28247	.351	.31953	.400	.35182	.449	.37758
.254	.24260	.303	.28326	.352	.32024	.401	.35242	.450	.37801
.255	.24346	.304	.28406	.353	.32095	.402	.35301	.451	.37845
.256	.24432	.305	.28485	.354	.32166	.403	.35361	.452	.37888
.257	.24518	.306	.28564	.355	.32237	.404	.35420	.453	.37930
.258	.24604	.307	.28643	.356	.32307	.405	.35479	.454	.37972
.259	.24690	.308	.28722	.357	.32377	.406	.35537	.455	.38014
.260	.24775	.309	.28801	.358	.32447	.407	.35595	.456	.38055
.261	.24860	.310	.28879	.359	.32517	.408	.35653	.457	.38096
.262	.24946	.311	.28958	.360	.32586	.409	.35711	.458	.38136
.263	.25021	.312	.29036	.361	.32655	.410	.35768	.459	.38176
.264	.25116	.313	.29114	.362	.32725	.411	.35825	.460	.38216

V'rs'd Sine	Area of Segment	V'rs'd Sine	Area of Segment	V'rs'd Sine	Area of Segment	V'rs'd Sine	Area of Segment	V'rs'd Sine	Area of Segment
.461	.38255	.469	.38549	.477	.38808	.485	.39026	.493	.39120
.462	.38293	.470	.38583	.478	.38837	.486	.39050	.494	.39208
.463	.38331	.471	.38617	.479	.38866	.487	.39073	.495	.39222
.464	.38369	.472	.38650	.480	.38895	.488	.39095	.496	.39236
.465	.38406	.473	.38683	.481	.38922	.489	.39116	.497	.39248
.466	.38442	.474	.38715	.482	.38949	.490	.39137	.498	.39258
.467	.38478	.475	.38746	.483	.38975	.491	.39156	.499	.39265
.468	.38514	.476	.38777	.484	.39001	.492	.39174	.500	.39269

USE OF THE ABOVE TABLE

To find the area of a circular zone.

Rule 1.—When the zone is less than a semicircle, divide the height by the longest chord, and seek the quotient in the column of versed sines. Take out the corresponding area, in the next column on the right hand, and multiply it by the square of the longest chord; the product will be the area of the zone.

Example.—Required the area of a zone, whose longest chord is 50, and height 15.

$15 \div 50 = .300$; and .300, as per table=.28087.

Hence, $.28087 \times 50^2 = 702.19$, the area of the zone.

Rule 2.—*When the zone is greater than a semicircle*, take the height on each side of the diameter of the circle, and find, by Rule 1, their respective areas; the areas of these two portions, added together, will be the area of the zone.

Example.—Required the area of a zone, the diameter of the circle being 50, and the height of the zone on each side of the line which passes through the diameter of the circle 20 and 15, respectively.

$20 \div 50 = .400$; .400, as per table=.35182; and $.35182 \times 50^2 = 879.56$.

$15 \div 50 = .300$; .300, as per table=.28087; and $.28087 \times 50^2 = 702.19$.

Hence, $879.56 + 702.19 = 1581.75$.

TABLE OF THE PROPORTIONS OF THE LENGTHS OF CIRCULAR ARCS

H'ght of Arc	Length of Arc	H'ght of Arc	Length of Arc	H'ght of Arc	Length of Arc	H'ght of Arc	Length of Arc	H'ght of Arc	Length of Arc
.100	1.0265	.144	1.0544	.188	1.0917	.232	1 1379	.276	1.1921
.101	1.0270	.145	1.0552	.189	1.0927	.233	1.1390	.277	1 1934
.102	1.0275	.146	1.0559	.190	1.0936	.234	1.1402	.278	1.1948
.103	1.0281	.147	1.0567	.191	1.0946	.235	1.1414	.279	1.1961
.104	1.0286	.148	1.0574	.192	1.0956	.236	1.1425	.280	1.1974
.105	1.0291	.149	1.0582	.193	1.0965	.237	1.1436	.281	1.1989
.106	1.0297	.150	1.0590	.194	1.0975	.238	1.1448	.282	1.2001
.107	1.0303	.151	1.0597	.195	1.0985	.239	1.1460	.283	1.2015
.108	1.0308	.152	1.0605	.196	1.0995	.240	1.1471	.284	1.2028
.109	1.0314	.153	1.0613	.197	1.1005	.241	1.1483	.285	1.2042
.110	1.0320	.154	1.0621	.198	1.1015	.242	1.1495	.286	1.2056
.111	1.0325	.155	1.0629	.199	1.1025	.243	1.1507	.287	1.2070
.112	1.0331	.156	1.0637	.200	1.1035	.244	1.1519	.288	1.2083
.113	1.0337	.157	1.0645	.201	1.1045	.245	1.1531	.289	1.2097
.114	1.0343	.158	1.0653	.202	1.1055	.246	1.1543	.290	1.2120
.115	1.0349	.159	1.0661	.203	1.1065	.247	1.1555	.291	1.2124
.116	1.0355	.160	1.0669	.204	1.1075	.248	1.1567	.292	1.2138
.117	1.0361	.161	1.0678	.205	1.1085	.249	1.1579	.293	1.2152
.118	1.0367	.162	1.0686	.206	1.1096	.250	1.1591	.294	1.2166
.119	1.0373	.163	1.0694	.207	1.1006	.251	1.1603	.295	1.2179
.120	1.0380	.164	1.0703	.208	1.1117	.252	1.1616	.296	1.2193
.121	1.0386	.165	1.0711	.209	1.1127	.253	1.1628	.297	1.2206
.122	1.0392	.166	1.0719	.210	1.1137	.254	1.1640	.298	1.2220
.123	1.0399	.167	1.0728	.211	1.1148	.255	1.1653	.299	1.2235
.124	1.0405	.168	1.0737	.212	1.1158	.256	1.1665	.300	1.2250
.125	1.0412	.169	1.0745	.213	1.1169	.257	1.1677	.301	1.2264
.126	1.0418	.170	1.0754	.214	1.1180	.258	1.1690	.302	1.2278
.127	1.0425	.171	1.0762	.215	1.1190	.259	1.1702	.303	1.2292
.128	1.0431	.172	1.0771	.216	1.1201	.260	1.1715	.304	1.2306
.129	1.0438	.173	1.0780	.217	1.1212	.261	1.1728	.305	1.2321
.130	1.0445	.174	1.0789	.218	1.1223	.262	1.1740	.306	1.2335
.131	1.0452	.175	1.0798	.219	1.1233	.263	1.1753	.307	1.2349
.132	1.0458	.176	1.0807	.220	1.1245	.264	1.1766	.308	1.2364
.133	1.0465	.177	1.0816	.221	1.1256	.265	1.1778	.309	1.2378
.134	1.0472	.178	1.0825	.222	1.1266	.266	1.1791	.310	1.2393
.135	1.0479	.179	1.0834	.223	1.1277	.267	1.1804	.311	1.2407
.136	1.0486	.180	1.0843	.224	1.1289	.268	1.1816	.312	1.2422
.137	1.0493	.181	1.0852	.225	1.1300	.269	1.1829	.313	1.2436
.138	1.0500	.182	1.0861	.226	1.1311	.270	1.1843	.314	1.2451
.139	1.0508	.183	1.0870	.227	1.1322	.271	1.1856	.315	1.2465
.140	1.0515	.184	1.0880	.228	1.1333	.272	1.1869	.316	1.2480
.141	1.0522	.185	1.0889	.229	1.1344	.273	1.1882	.317	1.2495
.142	1.0529	.186	1.0898	.230	1.1356	.274	1.1897	.318	1.2510
.143	1.0537	.187	1.0908	.231	1.1367	.275	1.1908	.319	1.2524

H'ght of Arc	Length of Arc	H'ght of Arc	Length of Arc	H'ght of Arc	Length of Arc	H'ght of Arc	Length of Arc	H'ght of Arc	Length of Arc
.320	1.2539	.357	1.3112	.393	1.3711	.429	1.4349	.465	1.5022
.321	1.2554	.358	1.3128	.394	1.3728	.430	1.4367	.466	1.5042
.322	1.2569	.359	1.3144	.395	1.3746	.431	1.4386	.467	1.5061
.323	1.2584	.360	1.3160	.396	1.3763	.432	1.4404	.468	1.5080
.324	1.2599	.361	1.3176	.397	1.3780	.433	1.4422	.469	1.5099
.325	1.2614	.362	1.3192	.398	1.3797	.434	1.4441	.470	1.5119
.326	1.2629	.363	1.3209	.399	1.3815	.435	1.4459	.471	1.5138
.327	1.2644	.364	1.3225	.400	1.3832	.436	1.4477	.472	1.5157
.328	1.2659	.365	1.3241	.401	1.3850	.437	1.4496	.473	1.5176
.329	1.2674	.366	1.3258	.402	1.3867	.438	1.4514	.474	1.5196
.330	1.2689	.367	1.3274	.403	1.3885	.439	1.4533	.475	1.5215
.331	1.2704	.368	1.3291	.404	1.3902	.440	1.4551	.476	1.5235
.332	1.2720	.369	1.3307	.405	1.3920	.441	1.4570	.477	1.5254
.333	1.2735	.370	1.3323	.406	1.3937	.442	1.4588	.478	1.5274
.334	1.2750	.371	1.3340	.407	1.3955	.443	1.4607	.479	1.5293
.335	1.2766	.372	1.3356	.408	1.3972	.444	1.4626	.480	1 5313
.336	1.2781	.373	1.3373	.409	1.3990	.445	1.4644	.481	1.5332
.337	1.2786	.374	1.3390	.410	1.4008	.446	1.4663	.482	1.5352
.338	1.2812	.375	1.3406	.411	1.4025	.447	1.4682	.483	1.5371
.339	1.2827	.376	1.3423	.412	1.4043	.448	1.4700	.484	1.5391
.340	1.2843	.377	1.3440	.413	1.4061	.449	1.4719	.485	1.5411
.341	1.2858	.378	1.3456	.414	1.4079	.450	1.4738	.486	1.5430
.342	1.2874	.379	1.3473	.415	1.4097	.451	1.4757	.487	1.5450
.343	1.2890	.380	1.3490	.416	1.4115	.452	1.4775	.488	1.5470
.344	1.2905	.381	1.3507	.417	1.4132	.453	1.4794	.489	1.5489
.345	1.2921	.382	1.3524	.418	1.4150	.454	1.4813	.490	1.5509
.346	1.2937	.383	1.3541	.419	1.4168	.455	1.4832	.491	1.5529
.347	1.2952	.384	1.3558	.420	1.4186	.456	1.4851	.492	1.5549
.348	1.2968	.385	1.3574	.421	1.4204	.457	1.4870	.493	1.5569
.349	1.2984	.386	1.3591	.422	1.4222	.458	1.4889	.494	1.5585
.350	1.3000	.387	1.3608	.423	1.4240	.459	1.4908	.495	1.5608
.351	1.3016	.388	1.3625	.424	1.4258	.460	1.4927	.496	1.5628
.352	1.3032	.389	1.3643	.425	1.4276	.461	1.4946	.497	1.5648
.353	1.3047	.390	1.3660	.426	1.4295	.462	1.4965	.498	1.5668
.354	1.3063	.391	1.3677	.427	1.4313	.463	1.4984	.499	1.5688
.355	1.3079	.392	1.3694	.428	1.4331	.464	1.5003	.500	1.5708
.356	1.3095								

To find the length of an arc of a circle by the foregoing table.

Rule.—Divide the height by the base, and the quotient will be the height of an arc, of which the base is unity. Seek in the table for a number corresponding to the quotient, and take the length of that height from the next right-hand column. Multiply the number, thus found, by the base of the arc, and the product will be the length of the arc or curve required.

Example.—The profiles of the intradoses of the arches of a bridge are each a semi-ellipse; the span of the middle arch is 150 feet, and the height 38 feet: required the length of the curve.

38+150=.253, and .253, as per table=1.1628.
Hence 1.1628×150=174.4200, the length required.

TABLE OF THE PROPORTIONS OF THE LENGTHS OF SEMI-ELLIPTIC ARCS

H'ght of Arc	Length of Arc	H'ght of Arc	Length of Arc	H'ght of Arc	Length of Arc	H'ght of Arc	Length of Arc	H'ght of Arc	Length of Arc
.100	1.0416	.265	1.2306	.450	1.4931	.635	1.7850	.820	2.0971
.101	1.0426	.270	1.2371	.455	1.5008	.640	1.7931	.825	2.1060
.102	1.0436	.275	1.2436	.460	1.5084	.645	1.8013	.830	2.1148
.103	1.0446	.280	1.2501	.465	1.5161	.650	1.8094	.835	2.1237
.104	1.0456	.285	1.2567	.470	1.5238	.655	1.8176	.840	2.1326
.105	1.0466	.290	1.2634	.475	1.5316	.660	1.8258	.845	2.1416
.110	1.0516	.295	1.2700	.480	1.5394	.665	1.8340	.850	2.1505
.115	1.0567	.300	1.2767	.485	1.5472	.670	1.8423	.855	2.1595
.120	1.0618	.305	1.2834	.490	1.5550	.675	1.8505	.860	2.1685
.125	1.0669	.310	1.2901	.495	1.5629	.680	1.8587	.865	3.1775
.130	1.0720	.315	1.2960	.500	1.5709	.685	1.8670	.870	2.1866
.135	1.0773	.320	1.3038	.505	1.5785	.690	1.8753	.875	2.1956
.140	1.0825	.325	1.3106	.510	1.5863	.695	1.8836	.880	2.2047
.145	1.0879	.330	1.3175	.515	1.5941	.700	1.8919	.885	2.2139
.150	1.0933	.335	1.3244	.520	1.6019	.705	1.9002	.890	2.2230
.155	1.0989	.340	1.3313	.525	1.6097	.710	1.9085	.895	2.2322
.160	1.1045	.345	1.3383	.530	1.6175	.715	1.9169	.900	2.2414
.165	1.1106	.350	1.3454	.535	1.6253	.720	1.9253	.905	2.2506
.170	1.1157	.355	1.3525	.540	1.6331	.725	1.9337	.910	2.2597
.175	1.1213	.360	1.3597	.545	1.6400	.730	1.9422	.915	2.2689
.180	1.1270	.365	1.3669	.550	1.6488	.735	1.9506	.920	2.2780
.185	1.1327	.370	1.3741	.555	1.6567	.740	1.9599	.925	2.2872
.190	1.1384	.375	1.3815	.560	1.6646	.745	1.9675	.930	2.2964
.195	1.1442	.380	1.3888	.565	1.6725	750	1.9760	.935	2.3056
.200	1.1501	.385	1.3961	.570	1.6804	.755	1.9845	.940	2.3148
.205	1.1560	.390	1.4034	.575	1.6883	.760	1.9931	.945	2.3241
.210	1.1620	.395	1.4107	.580	1.6963	.765	2.0016	.950	2.3335
.215	1.1680	.400	1.4180	.585	1.7042	.770	2.0102	.955	2.3429
.220	1.1741	.405	1.4253	.590	1.7123	.775	2.0187	.960	2.3524
.225	1.1802	.410	1.4327	.595	1.7203	.780	2.0273	.965	2.3619
.230	1.1864	.415	1.4402	.600	1.7283	.785	2.0360	.970	2.3714
.235	1.1926	.420	1.4476	.605	1.7364	.790	2.0446	.975	2.3810
.240	1.1989	.425	1.4552	.610	1.7444	.795	2.0533	.980	2.3906
.245	1.2051	.430	1.4627	.615	1.7525	.800	2.0620	.985	2.4002
.250	1.2114	.435	1.4702	.620	1.7606	.805	2.0708	.990	2.4098
.255	1.2177	.440	1.4778	.625	1.7687	.810	2.0795	.995	2.4194
.260	1.2241	.445	1.4854	.630	1.7768	.815	2.0883	.1000	2.4291

To find the length of the curve of a right semi-ellipse.

Rule.—The rule for circular arcs in the preceding table is equally applicable here.

The two last tables are not entirely confined to works which may be carried into practice, but are useful in estimating, to a very minute degree of accuracy, the quantity of work which is to be executed from drawings to a scale.

As the tables, however, do not afford the means of finding the lengths of the curves of elliptic arcs, which are less than half of the entire figure, the following geometrical method is given to supply the defect.

To find the length of an elliptic curve, which is less than half the figure.

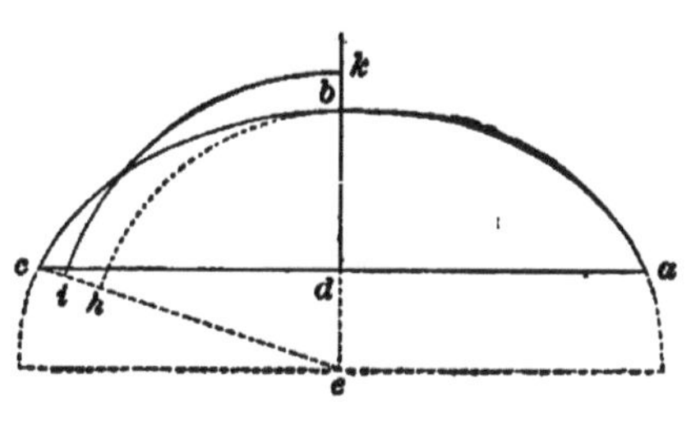

Let the curve, of which the length is required to be found, be *a b c*.

Produce the versed sine, *b d*, to meet the center of the curve in *e*. Draw the right line, *c e*, and from the center, *e*, with the distance, *e b*, describe an arc, *b h*. Bisect *c h* in *i*, and from the center, *e*, with the radius, *e i*, describe the arc, *i k*, meeting *e b* produced to *k*; then, *i k* is half the arc *a b c*.

NOTE *.—When the quotient is not given in the column of heights, divide the difference between the two nearest heights by .5; multiply the quotient by the excess of the height given, and the height in the table first above it, and add this sum to the tabular area of the least height.

Thus, if the height is 118,
.120, per table,=1.0618
.115, " " =1.0567
.0051 ÷ 5 = .00102 × (118—115) = .00306,
which, added to 1.0567=1.05976, the length for 118.

* Haswell.

OF SOLIDS BOUNDED BY PLANE SURFACES

The mensuration of solids is divided into two parts.

I. The mensuration of the surfaces of solids.

II. The mensuration of their solidities.

The *measure* of any solid body is the whole capacity or contents of that body, when considered under the triple dimensions of length, breadth, and thickness. A *cube*, whose side is one inch, one foot, or one yard, etc., is called the *measuring unit;* and the contents or solidity of any figure is computed by the number of those cubes contained in that figure.

DEFINITIONS

1. A *cube* is a right prism, bounded by six equal square faces, of which any two, opposite to each other, are parallel.

2. A *parallelopiped* is a prism bounded by six quadrilateral planes, every opposite two of which are equal and parallel.

3. A *prism* is a solid, whose ends are parallel, similar, and equal, and the sides connecting these are parallelograms.

4. A *pyramid* is a solid, whose base is any plane figure, and whose sides are triangles, having all their vertices meeting together in a point above the base, called the *vertex* of the pyramid.

5. A *frustum* or *trunk* of a pyramid is a portion of the solid that remains after any part has been cut off parallel to the base.

6. A *wedge* is a solid of five sides, two of which are rhomboidal, and meet in an edge, a rectangular base, and two triangular ends.

7. A *prismoid* is a solid, whose ends or bases are parallel, but not similar, and whose sides are quadrilateral.

OF CUBES AND PARALLELOPIPEDS

Problem I.—To find the lateral surface of a prism.

Rule.—Multiply the perimeter of the base into the altitude, and the product will be the convex, or lateral surface. When the *entire* surface of the prism is reqnired, add to the convex surface the area of the bases.

Example.—Required the lateral surface of a prism whose base is a regular hexagon, and whose sides are each 2 feet 3 inches, the height being 11 feet?

2 ft. 3 in =27 in. and 27×6=perimeter of the base.
11 ft. =132 inches=height.
Then, 132×162=21384 square inches.
21384÷144=148.50 sq. ft. Ans.

Problem II.—To find the solidity of a cube or right prism.

Rule.—Multiply the area of the base by the perpendicular height, and the product will be the solid contents.

NOTE.—The capacity of a vessel, in gallons or bushels, of any given dimensions, may be readily ascertained by calculating its contents in *inches*, and then dividing the contents by the number of cubic inches in one gallon or bushel.

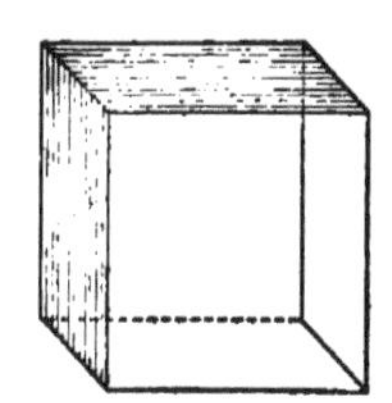

Examples.—1. Required the number of *ale* gallons there are in a *cistern* which is 6 feet 8 inches deep, and whose base is 5 feet 4 inches square?

6 ft. 8 in.=80 in.
5 ft. 4 in.=64 in.
Then, $\overline{64}^2$=4096, and 4096×80=327680=solidity in inches.
And 327680÷282=1162 gal.

2. What is the solidity of a *prism* of granite, 9 feet 2 inches long, and 16 by 12 inches side dimension, and

what will be its weight, reckoning 169 lbs. to the cubic foot?

9 ft. 2 in.=110 in.=length. | 192×110=21120=solidity in in.
16×12=192 in.=area of base | 21120÷1728=12.22 cubic ft. Ans.
12.22×169=2065 lbs. Ans.

OF PYRAMIDS

Problem III.—To find the lateral surface of a regular pyramid.

Rule.—Multiply the perimeter of the base by the slant height, and half the product will be the surface. If the whole surface be required, add to this the area of the base.

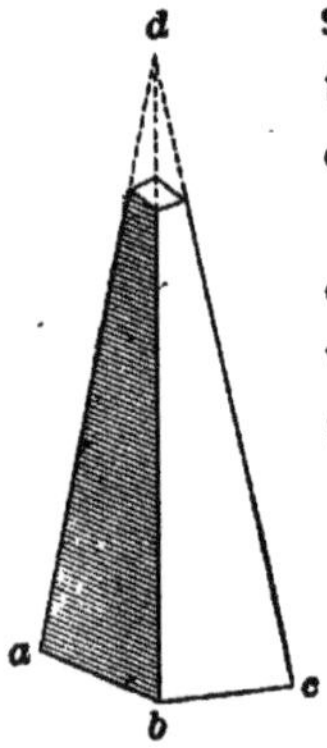

Example.—What is the lateral surface of a regular triangular pyramid, *a b c*, whose slant height, *d a*, is 20 feet, and the sides of whose base are each 8 feet?

8×3=24=perimeter of the base.
20=slant height.
2)480
240=lateral surface.

Problem IV.—To find the lateral surface of the frustum of a regular pyramid.

Rule.—Multiply the perimeters of the two ends by the slant height of the frustum, and half the product will be the surface required. To this add the surface of the two ends when the entire surface is required.

Example.—What is the lateral surface of the frustum of a regular octagonal pyramid, *A B C D*, whose slant height, *a A*, is 42 feet, and the sides of the lower base, *D C*, 5 feet each, and of the upper base, *a b*, 3 feet each?

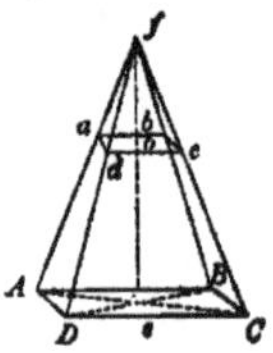

First, 5×8=40=perimeter of lower base.
3×8=24= " upper "
64=sum of the two ends.
Then, 64=42÷2=1344=area of lateral surface.

Problem V.—To find the solidity of a pyramid.

Rule.—Find the area of the base, and multiply that area by ⅓ of the height.

NOTE.—This rule follows from that of the *prism*, because any pyramid is ⅓ of a prism of the same base and altitude. It is manifest, therefore, that the solidity of a pyramid, whether right or oblique, is equal to the product of the area of the base into ⅓ of the perpendicular height.

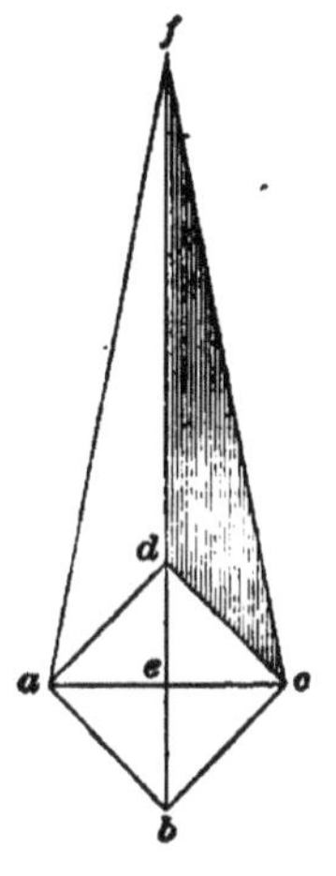

Example.—What is the solidity of a square pyramid, *a b c d*, the sides of whose base are each 30 feet, and its perpendicular height, *e f*, 25 feet?

First, $30 \times 30 = 900$ = area of the base.

$$\begin{array}{r} 25 \div 3 = \quad 8\tfrac{1}{3} \\ \hline 7200 \\ 300 \\ \hline 7500 \end{array}$$

7500 = solidity.

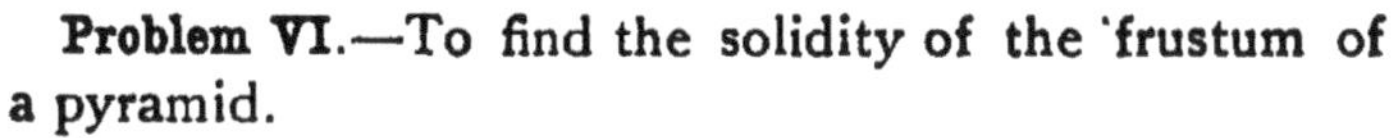

Problem VI.—To find the solidity of the frustum of a pyramid.

Rule.—To the areas of the two ends of the frustum, add the square root of their product; and this sum, multiplied by ⅓ of the perpendicular height, will give the solid contents.

NOTE.—This rule holds equally true to a pyramid of any form. For the solidities of pyramids are equal when they have equal heights and bases, whatever be the figure of their bases.

Example.—What is the cubic or solid contents of the frustum of a marble pyramid, whose lower base, *a b c d*, is 20 inches square, and upper base, *e f*, 14 inches, and whose height, *h g*, is 8 feet 4 inches? And what will be its weight, reckoning 169 lbs. to the cubic foot?

$\overline{20}^2$=400=area of lower base. 8 ft. 4=100

$\overline{14}^2$=196= " upper " 100÷3=33⅓=⅓ of height.

$\overline{596}$=sum of areas. Then, $\sqrt{400 \times 196}$=280.

And, 596+280×33⅓=29200.

2920÷1728=16.9 cubic feet. Ans.

To find the weight, 16.9×169=2856 lbs. Ans.

NOTE.—By this rule, marble cutters can easily determine the solidity and weight of any piece of marble, such as shafts of monuments, slabs, etc., by reference to the Table of Specific Gravities, for a multiplier for the weight of a cubic foot or inch.

OF WEDGES AND PRISMOIDS

Problem VII.—To find the solidity of a wedge.

Rule.—To the length of the edge of the wedge add twice the length of the base.

Then multiply this sum by the height of the wedge and the breadth of the base, and ⅙ of the product will be the solid contents.

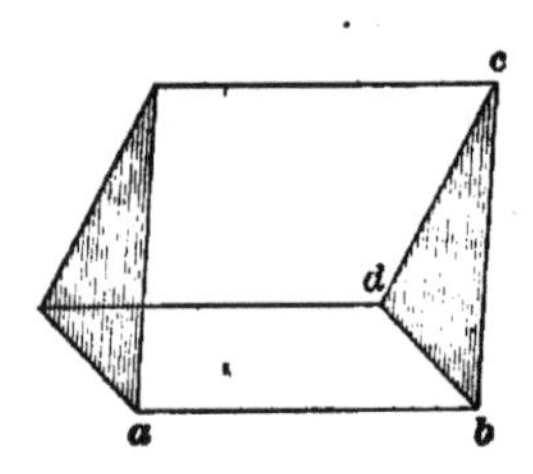

Example.—Required the solidity of a wedge whose base, *a b*, is 27 feet, *b d*, 8 feet, and whose edge, *c b*, is 36 feet, and the perpendicular height 22 feet?

First, 36=length of edge.

54=twice the length of the base.

$\overline{90}$×22×8÷6=2660 cubic ft.

Problem VIII.—To find the solidity of a rectangular prismoid.

Rule.—To the sum of the areas of the two ends, *a b c*, *d e f*, add four times the area of a section, *g h*, parallel to and equally distant from the parallel ends, and this sum, multiplied by ⅙ of the height, will give the solidity.

Example.—What is the solidity of a rectangular prismoid, *a b c d*, the length and breadth of one end being 14 by 12 inches and the other 6 by 4 inches, and the perpendicular 30 feet 6 inches?

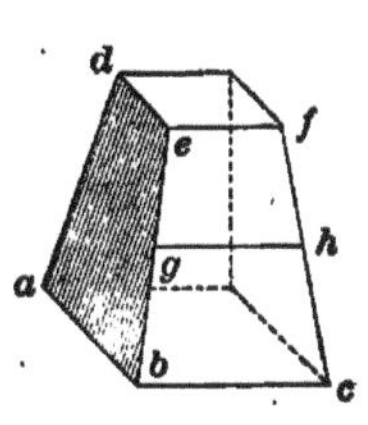

First, 14×12=168=area of lower base.
6× 4= 24= " upper "
192

14+6+2=10 } length and breadth
12+4+2= 8 } of middle section
80
4
320=area of 4 times middle section.

Then,
192
320
512
61=⅙ height
512
3072
31232

And 31232÷1728=18.074 cubic ft. Ans.

OF THE CYLINDER, CONE, AND SPHERE

DEFINITIONS

1. A *cylinder* is a solid, having equal and parallel circles for its ends, and is described by the revolution of a rectangle about one of its sides.

2. A *cone* is a solid body, of a true taper from the base to a point, which is called the vertex, and has a circle for its base.

3. A *frustum* of a cone is what remains after a portion is cut off by a plane, parallel to the base.

4. A *conoid* is a solid, generated by the revolving of a parabola or hyperbola around its axes.

5. A *spheroid* is a solid, generated by the revolution of an ellipse about either of its axes.

6. A *sphere* is a solid, terminated by a curved surface, all the points of which are equally distant from a point within, called the center. A sphere may be described by the revolution of a semicircle about a diameter.

7. A *radius* of a sphere is a line drawn from the center to any part of the surface; as,

8. The *diameter* of a sphere is a line drawn through the center, and terminated at both ends by the surface. All diameters of a sphere are equal to each other, and each is double the radius.

9. A *segment* of a sphere is a portion of the sphere cut off by any plane. This plane is called the *base* of the segment. The *height* of a segment is the distance from the middle of its base to the convex surface.

10. A *zone* is a portion of the surface of a sphere, included between two parallel planes, which form its bases. If the bases are equally distant from the center, it is called the *middle zone*. The *height* of a zone is the perpendicular distance between the two planes which form its bases.

11. A *cylindrical ring* is a solid, formed by bending a cylinder, as a cylindrical bar of iron, until the two ends meet each other.

12. A *parabola* is a section of a cone when cut by a plane parallel to its sides.

13. A *hyperbola* is the section of a cone when cut by a plane, making a greater angle with the base than the side of a cone makes.

14. The *transverse axis* is the longest straight line that can be drawn in an ellipse.

15. The *conjugate axis* is a line drawn through the center, at right angles to the transverse axis.

16. An *abscissa* is a part of any diameter contained between its vertex and an ordinate.

17. The *focus* is the point in the axis where the ordinate is equal to half the perimeter.

Problem I.—To find the convex surface of a cylinder.

Rule.—Multiply the circumference of the base by the

length of the cylinder, and the product will be the convex surface required. To this add the areas of the two ends when the entire surface is required.

Example.—What is the convex surface of a right cylinder, whose length is 23 feet, and the diameter of its base 3 feet?

$3 \times 3.14159 = 9.42477$
Then, $9.42477 \times 23 = 216.76971$ = surface.

Problem II.—To find the solidity of a cylinder.

Rule.—Multiply the area of the base by the height, and the product will give the solid contents.

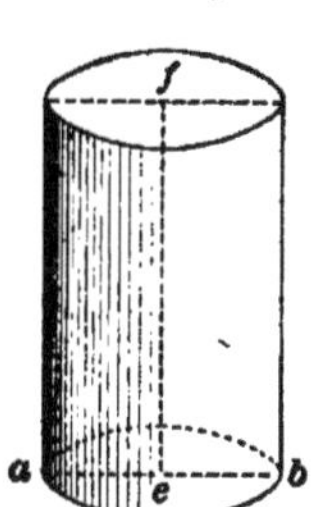

Examples.—1. What is the solidity of a cylinder, the diameter, *a b*, of whose base is 16 feet, and its height, *e f*, 28 feet?

First, find the area of the base by $\overline{16}^2 = 256$.
Then, $256 \times .7854 = 201.0624$ = area of the base.
Then, $201.0624 \times 28 = 5629.7472$ = solid contents.

2. The Winchester bushel is a hollow cylinder, 18½ inches in diameter and 8 inches deep: what is its capacity?

First, the area of the base $= \overline{18.5}^2 \times .7854 = 268.8025$.
Then, $268.8025 \times 8 = 2150.42$ = capacity in cubic inches.

NOTE.—By this rule, every sealer of weights and measures may determine the exact capacity of any *measure* submitted to his inspection. And so any one may test the accuracy of any measure, whether dry or liquid, by reducing its capacity to cubic inches, and dividing by the number of cubic inches contained in such measure. The divisor for any measure may be found in the Table of Weights and Measures.

3. How many gallons of oil will a can of a cylindrical form hold, whose diameter is 28⅝ inches, and whose height is 4 feet 3 inches?

Area of the base by the Tables of Areas of Circles=643.54;
and 643.54×51÷221.1841=48.39 gallons.
1 gallon=221.184 cubic inches.

Problem III.—To find the convex surface of a cone.

Rule.—Multiply the perimeter of the base by the slant height, and ½ the product will be the surface; to which add the area of the base when the entire surface is required.

Example.—The diameter of the base of a right cone, *a b*, is 3 feet, and the slant height, *c a*, is 15 feet: what is the convex surface?

First, 3×3.14159=9.42477=circum. of base.
Then, 9.42477×15÷2=70.686 sq. ft.

Problem IV.—To find the solidity of a cone.

Rule.—Multiply the area of the base by ⅓ of the height, and the product will be the solidity.

Example.—What is the solidity of a right cone, whose perpendicular height, *c d*, is 10½ feet, and the circumference of the base is 9 feet?

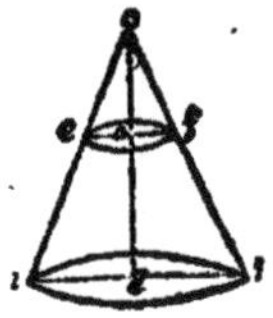

We here multiply the area of the base by ⅓ of the height, and the product is the solidity.

First, 9^2=81, and 10½÷3=3½=⅓ height.
Now, 81×.7854=63.6174, area of base.
Then, 63.6174×3½=222.6609. Ans.

Problem V.—To find the surface of a frustum of a cone.

Rule.—Add together the circumferences of the two ends, and multiply the sum by ½ the slant of the frustum; the product will be the convex surface: to which add the areas of the two bases when the entire surface is required.

NOTE.—This rule is precisely the same as that for a *frustum* of a pyramid, and if a cone be considered as a pyramid of an infinite number of sides, it is equally applicable to the measurement of the *frustum* of a cone.

Example.—What is the convex surface of the frustum of a cone, the circumference of the greater base, *a b*, being 30 feet, and of the smaller, *e f*, 10 feet, the slant height, *c a*, being 20 feet?

30+10=40=circum. of two ends. 10=½ slant height.
40×10=400=convex surface.

Problem VI.—To find the solidity of the frustum of a cone.

Rule.—Add to the areas of the two ends of the frustum the *square root of their product.* Then multiply this sum by ⅓ of the perpendicular height, and the product will be the solidity.

NOTE.—If a *cone* and a *pyramid* have equal bases and altitudes, they are equal in their solidity. Consequently, the rule already given for the *frustum* of a *pyramid* is equally applicable to the frustum of a cone.

Example.—How many gallons of ale are contained in a cistern in the form of a conic frustum, *a b e f*, if the larger diameter, *a b*, be 9 feet, and the smaller diameter, *e f*, 7 feet, and the depth, *c o d*, 9 feet?

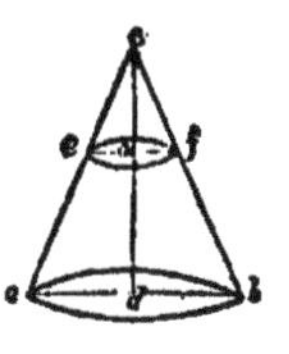

9^2=81 and { 81× 7854=63.61=area of lower base.
7^2=49 and { 49×.7854=38.48= " upper "
102.09

Then, 63.61×38.48=2447.71 | 102.09+49.46=151.55.
$\sqrt{2447.71}$=49.46 | 151.55×3=454.65 cubic feet.

454.65×1728=785635 cubic inches.
785635÷282=2785 gal. Ans.

OF SPHERES

Problem VII.—To find the surface of a sphere or globe.

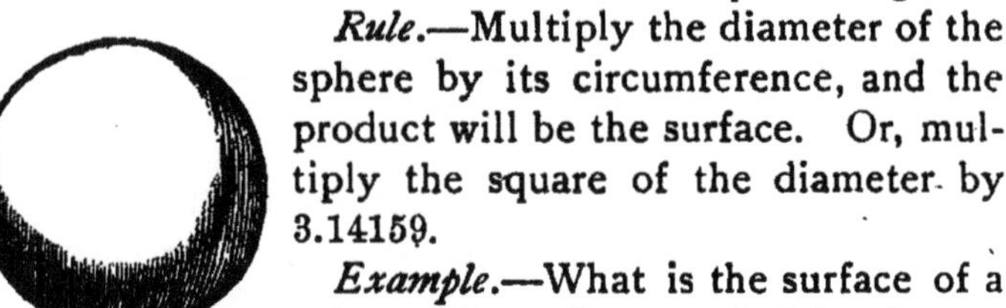

Rule.—Multiply the diameter of the sphere by its circumference, and the product will be the surface. Or, multiply the square of the diameter by 3.14159.

Example.—What is the surface of a sphere whose diameter is 7 feet?

First, $7 \times 3.14159 = 21.99113$ = circumference.
Then, $21.99113 \times 7 = 153.93791$ sq. ft. = surface.

Problem VIII.—To find the convex surface of a spherical zone or segment.

Rule.—Multiply the height of the zone or segment by the whole circumference of the sphere of which it is a part, and the product will be the convex surface.

Example.—If the axis of a sphere be 42 inches, what is the convex surface of a segment or zone, *a b d*, whose height, *c d*, is 9 inches?

First, $42 \times 3.14159 = 131.9468$ = circumference.
9 = height.
1187.5212 = surface in square inches.

Problem IX.—To find the solidity of a sphere or globe.

Rule.—Multiply the cube of the diameter, *c e*, by the decimal .5236. Or, multiply the square of the diameter by the circumference, and ⅙ of the product will be the contents.

Example.—What is the solidity of a globe whose diameter, *c e*, is 12 inches?

$\overline{12}^2 \times 3.14159 = 452.38996$ = surface of the sphere.
Then, $452.38996 \times 12 \div 6 = 904.78$ = solidity.
Or thus: $\overline{12}^3 = 1728$ = cube of the diameter.
And $1728 \times .5236 = 904.78$ = solid contents.

Problem X.—To find the solidity of a spherical segment.

Rule.—To three times the square of the radius, *a b*, of its base, add the square of its height, *b c;* then multiply the sum by the height, and the product by .5236, for the contents.

Example.—What is the solidity of the segment, *a d c* (of the sphere *e c*), whose height, *b c*, is 8 feet, and the diameter of whose base, *a d*, is 14 feet?

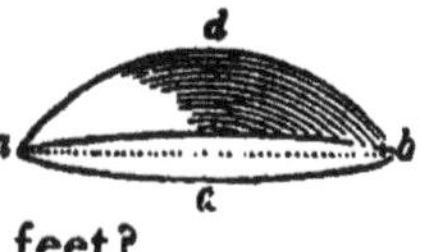

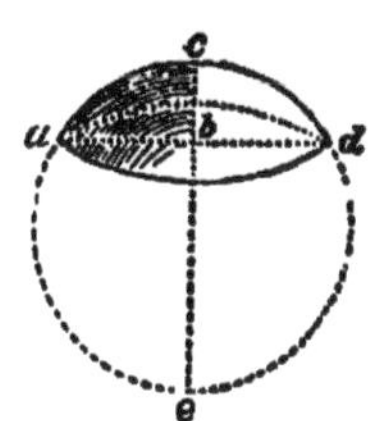

$7^2 = 49 \times 3 = 147$
$8^2 = \quad 64$
$\overline{211} \times 8 = 1688 \times .5236 = 883.836$. **Ans.**

NOTE.—The solidity of a spherical segment is frequently required when the radius of its base is not given; but if the *diameter* of the sphere and the height of the segment be known, the solidity may be easily found by the following:

Rule.—From three times the diameter of the sphere, subtract twice the height of the segment; then multiply the remainder by the square of the height, and the product by the decimal .5236.

OF SPHEROIDS

Problem XI.—To find the solidity of a spheroid.

Rule.—Multiply the square of the revolving axis by the fixed axis: and the product, multiplied by .5236, will give the solidity.

Example.—What is the solidity of an oblong spheroid, whose longer axis, *a b*, is 30, and the shorter, *c d*, 20, the revolving axis being *c d?*

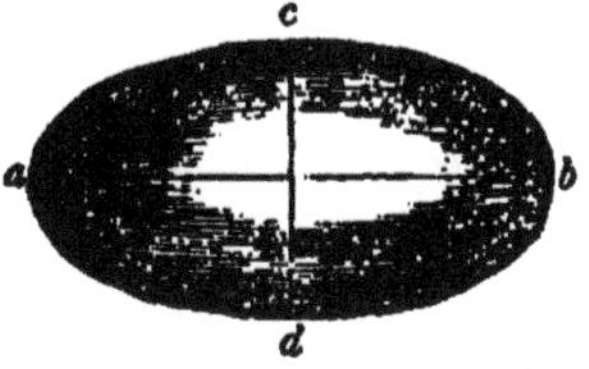

$\overline{20}^2 \times 30 = 12000$
Then, $12000 \times .5236 = 6283.2$. **Ans.**

NOTE.—If the generating ellipse revolves about its major axis, the spheroid is *prolate* or oblong; if about its minor axis, the spheroid is *oblate*.

OF PARABOLIC CONOIDS AND SPINDLES

Problem XII.—To find the solidity of a parabolic conoid.

Rule.—Multiply the square of the diameter of the base by the altitude, and the product by .3927 (which is ½ of .7854), and it will give the contents.

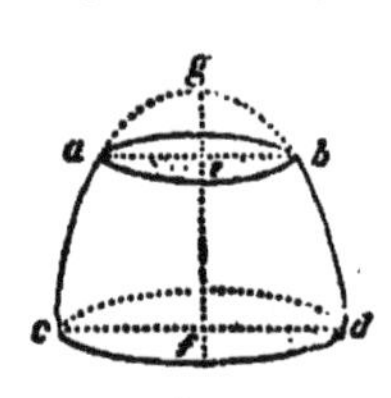

Example.—What is the solidity of a parabolic conoid, whose height, *f g*, is 60, and the diameter, *c d*, of its base 100 inches?

$\overline{100}^2=10000$

And $10000\times60\times.3927=235620$. Ans.

Problem XIII.—To find the solidity of a frustum of a paraboloid.

Rule.—Multiply the sum of the squares of the diameters of the two ends, *a b* and *c d*, by the height of the frustum, *e f*, and the product by .3927 (which is ½ of .7854), and it will give the contents.

Example.—What is the solidity of the *frustum* of a paraboloid, *a b c d*, whose diameter, *c d*, is 54, *a b*, 28, and height, *f e*, 18 inches?

$\overline{54}^2=2916$. Then, $3700\times18\times.3927=26153.82$. Ans.
$\overline{28}^2=\underline{\ 784}$
3700

Problem XIV.—To find the solidity of a parabolic spindle.

Rule.—Multiply the square of the middle diameter, *c d*, by the length of the spindle, *l m*, and the product by .41888 (which is $\frac{8}{15}$ of .7854), and it will give the solidity.

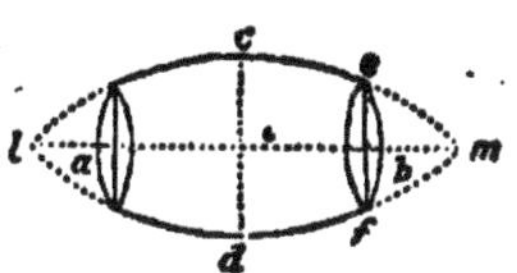

Example.—Required the solidity of the parabolic spindle, *l m*, *c d*, whose length, *l m*, is 100, and diameter, *c d*, 40.

$\overline{40}^2=1600.$

And $1600\times100\times.41888=67020.8.$ **Ans**

Problem XV.—To find the solidity of the middle frustum of a parabolic spindle.

Rule.—Add together 8 times the square of the greatest diameter, *c d*, 3 times the square of the least diameter, *f e*, and 4 times the product of these two diameters; multiply the sum by the length, *a b*, and the product by .05236 (which is $\frac{1}{60}$ of 3.1416); this will give the solidity.

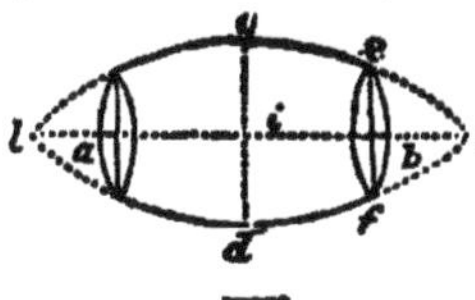

Example.—What is the solidity of the frustum of a parabolic spindle, whose dimensions are as follows: *a b*, 60, *c d*, 40, *f e*, 30 inches?

$$\begin{array}{rr}
\overline{40}^2= & 1600 \\
 & 8 \\
\hline
 & 12800 \\
\overline{30}^2=900\times3= & 2700 \\
\hline
 & 15500 \\
30\times40\times4= & 4800 \\
\hline
 & 20300\times60\times.05236=63774.48.\ \textbf{Ans.}
\end{array}$$

OF HYPERBOLOIDS AND HYPERBOLIC CONOIDS

Problem XVI.—To find the solidity of a hyperboloid.

Rule.—To the square of the radius of the base, *a s*, add the square of the middle diameter, *m r;* multiply this sum by the height, *s f*, and the product by .5236, and it will give the solidity.

Example.—What is the solidity of a hyperboloid, *a b f*, whose base, *a b*, is 40 inches, and height, *s f*, 30 inches; and whose middle diameter, *m r*, is 30 inches?

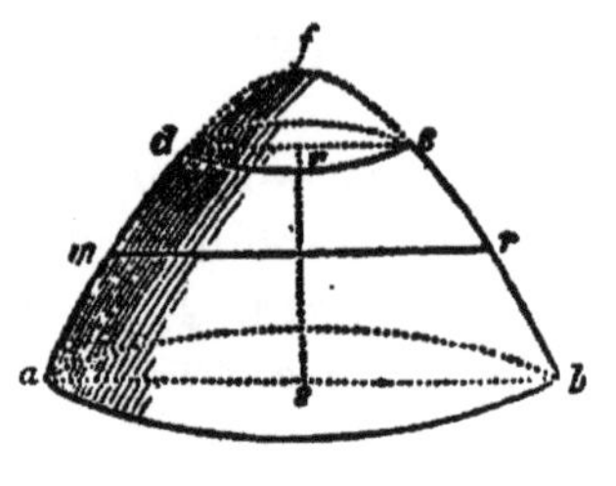

$\overline{20}^2=400$
$\overline{30}^2=900$
$\overline{1300}$ And 1300×30× 5236÷1728=11.817 cubic feet.

Problem XVII.—To find the solidity of the frustum of a hyperbolic conoid.

(See the foregoing figure.)

Rule.—Add together the squares of the greatest and least semidiameters, *a s* and *d r*, and the square of the whole diameter, *m r*, in the middle of the two; multiply this sum by the height, *r s*, and the product by .5236, and it will give the solidity.

Example.—Required the solidity of the frustum of a hyperbola, *a b d e*, whose semidiameter, *a s*, is 20 inches, and *d r*, 10 inches; the middle diameter, *m r*, 30 inches, and whose height is 20 inches?

$\overline{20}^2=400$
$\overline{10}^2=100$
$\overline{30}^2=900$
$\overline{1400}$ Then, 1400×20×.6236÷1728=8.426 cubic feet.

Problem XVIII.—To find the convex surface of a cylindrical ring.

Rule.—To the thickness of the ring, *a c*, add the inner diameter, then multiply this sum by the thickness, and the product by 9.8696 (which is the square of 3.1416), and it will give the convex surface required.

Example.—The thickness, *a c*, of a cylindrical ring is 4 inches, and the inner diameter, *c d*, is 14 inches; required the convex surface.

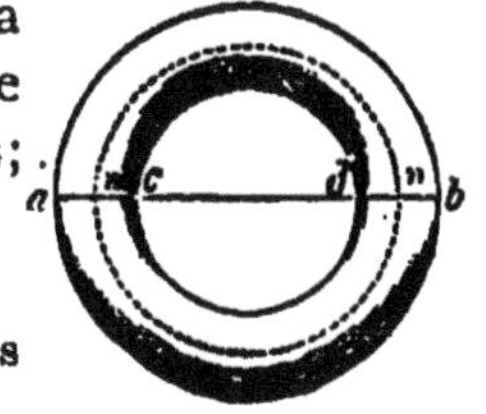

$ac+cd=4+14=18.$

Then, $18\times4\times9.8696=710.612$ square inches =convex surface.

Problem XIX.—To find the solidity of a cylindrical ring.

Rule.—To the thickness of the ring, *a c*, add the inner diameter, *c d;* then multiply the sum by the square of the thickness, and the product by 2.4674 (which is ¼ of the square of 3.1416), and it will give the solidity.

Example.—Required the solidity of an anchor ring, whose inner diameter is 8 inches, and thickness in metal 3 inches.

First, $3+8=11$

$3\times3=9$=square of thickness.

$99\times2.4674=244.2726$=solidity in inches.

GAUGING OF CASKS

Gauging is a practical art which does not admit of being treated in a very scientific manner.

Casks are not commonly constructed in exact conformity with any regular mathematical figure. By most writers on this subject, however, they are considered as nearly coinciding with one of the following forms:

1. } The middle frustum { of a spheroid,
2. } The middle frustum { of a parabolic spindle,
3. } The equal frustums { of a paraboloid,
4. } The equal frustums { of a cone,

and their contents in cubic inches may be found by the

rules in mensuration, for determining the solidity of these figures.

To find the contents of a cask by four dimensions.

Rule.—Add together the squares of the bung and head diameter, and the square of double the diameter, taken in the middle between the bung and head; multiply the sum by the length of the cask, and the product by .1309.

To find the contents of a cask in the form of the middle frustum of a spheroid.

Rule.—Add together the square of the head diameter and twice the square of the bung diameter; multiply the sum by ⅓ of the length, and the product by .00355, for a wine gallon of New York standard measure, or .0034 for old English gallons. If D and d = the two diameters, and l = the length, the capacity in inches $= (2D^2 \times d^2) \times \frac{1}{3}l \times .7854$. And by substituting .00355 for .7854, we have the capacity in wine gallons.

Example.—What is the capacity of a cask of the *second* form, whose length is 30 inches, its head diameter 18 inches, and its bung diameter 24?

$$\begin{aligned} \overline{18}^2 &= 324 \\ 2\times\overline{24}^2 &= 1152 \\ &\overline{1476} \\ \tfrac{1}{3}\text{ of }30 &= 10 \\ &\overline{14760}\times .00355 = 52.39 \text{ wine gallons. Ans.} \end{aligned}$$

To find the contents of a cask in the form of two equal frustums of a cone.

Rule.—Add together the square of the head diameter, the square of the bung diameter, and the *product* of the two diameters; multiply the sum by ⅓ of the length and the product by .00355 for New York wine gallons, or .0034 for old English gallons of 231 cubic inches.

Example.—What is the capacity of a cask whose dimensions are as follows: 30 inches long, head diameter 18 inches, and bung diameter 24 inches?

$$\overline{18}^2=324$$
$$\overline{24}^2=576$$
Product of 2 diam. $=432$
$$\overline{1332}\times 10=13320\times .00355=46.286$$
Or $(D^2+d^2+Dd)\times \frac{1}{3}l\times .00355.$

We are now in a position to commence work in earnest, and with this end in view we will start just as the workman starts, at the very beginning, which, in the case of a building, is the preparation of the site, the excavations, the drainage, the footings, the foundations, and so on, until the whole structure is finished; and I would like to remark before commencing that whatever method of estimating is started with, that method should be continued throughout the whole for that particular work. Sometimes, where there is any doubt as to the correctness of the result, it is a good way to finish up with one system, then to use another system, and if the two results are not wide apart, the estimate may be considered fairly correct. If, however, there is a big variation, the first estimate should again be gone over, and if the same result is obtained, or nearly the same, it may be considered fairly correct; it is well, however, to go over the second system again in order to find out where the discrepancy occurs. The price of accurate results is persistent effort.

In order to get at near approximation of the cost of work, the estimator, besides having a knowledge of the price of the various materials required, should be also conversant with the current price of labor, and to this end I give herewith the average price per hour of labor as now (1904) gathered from a number of labor circles throughout the whole country. These prices, however, are only given merely as guides, for they will vary with time and with locality; but in the absence of proper local data, they may be used with confidence. I give the price per hour of labor, as law or custom has not yet made the length of a legal day's labor.

AVERAGE RATES OF WAGES PER HOUR

General Laborer	from 15 to 20 cents
Stone Mason	from 40 to 50 cents
Excavator	from 18 to 22 cents
Bricklayer	from 35 to 45 cents
Carpenter	from 35 to 50 cents
Plasterer	from 35 to 40 cents
Slater	from 40 to 45 cents
Painter	from 22 to 35 cents
Plumber	from 45 to 55 cents
Roofer	from 35 to 45 cents

Other trades run in about the same proportion, so that, knowing the number of hours the work will require for completion, a fair estimate of the whole cost of the work may be arrived at.

A few of the things necessary to know in connection with estimating on excavation are the capacities of the tools and appliances required on the work, such as I give below.

An ordinary one-horse cart 6 feet long by 3½ feet wide and 2½ feet deep will hold 45 cubic feet, or 1⅔ cubic yards.

A regular builder's cart will hold 1 cubic yard.

A tip-wagon will hold, when heaped, 3 cubic yards.

A large wheelbarrow will hold 1/10 cubic yard.

A small wheelbarrow will hold 1/12 cubic yard.

A basket holds a bushel, or 1/21 cubic yard.

50 barrow loads make a good wagon load.

A stone wagon will carry from 2½ to 6 tons.

A double load of earth equals about 56 cubic feet.

A single load equals some 27 or 28 cubic feet.
A single, generally, is about 1 cubic yard.
A single, generally, is about 1 ton of stone, brick, etc.
500 bricks make a single load.
400 pressed bricks make a single load.
1,000 plain roofing tiles make a single load.
1,000 slates, counters, make a single load.
1,000 feet dressed lumber make a single load.
50 cubic feet of timber make a single load.
1 cubic yard of mixed mortar make a single load.
16 bushels of lime make a single load.

Earth in excavations weighs about as follows:

1 cubic yard of common earth, 2,400 pounds.
1 cubic yard of top-soil earth, 2,000 pounds.
1 cubic yard of clay earth, 2,700 pounds.
1 cubic yard dry sand earth, 2,700 pounds.
1 cubic yard wet sand earth, 3,000 pounds.
1 cubic yard of sandy loam earth, 2,400 pounds.
1 cubic yard of mud earth, 2,500 pounds.
1 cubic yard of gravel earth, 3,000 pounds.

1 cubic foot of cement concrete, 6 broken stones, 1 sand, 1 cement, weighs 130 pounds.

1 cubic foot of concrete, 6 broken bricks, 1 sand, and 1 cement, weighs 120 pounds.

1 cubic foot of concrete, 6 broken ballast, 1 sand, and 1 cement, weighs 140 pounds.

Increase in the bulk of earth, clay, etc., when excavated and thrown into a loose heap:

	BEFORE DIGGING	WHEN DUG
Earth and clay	1	1¼
Sand and gravel	1	1⅓
Broken stones	1	1⅓
Free stone	1	1¼
Rock generally	1	1½

STONE-WORK DRAIN TILES

125 pieces 2 feet long, 4-inch pipe, weigh 1 ton.
80 pieces 2 feet long, 6-inch pipe, weigh 1 ton.
42 pieces 2 feet long, 9-inch pipe, weigh 1 ton.
24 pieces 2 feet long, 12-inch pipe, weigh 1 ton.

COST OF LABOR IN PHILADELPHIA, BALTIMORE, CHICAGO, AND OTHER LARGE CENTERS, AT THE TIME OF COMPILATION OF THIS WORK

This is not to be considered reliable, but will answer when exact data are not at hand.

DESCRIPTION	MADE GROUND	COMMON GROUND	STIFF CLAY OR GRAVEL
	Cts.	Cts.	Cts.
Dig, throw out, and prepare for concrete, 12 inches deep, per super. yard	8	10	12
Digging and throwing out when more than 12 inches deep, including leveling, per cubic yard	15	20	25
Ditto in trenches, leveling, fixing, and removing, shoring and planking, not exceeding 6 feet deep, per cubic yard	20	26	32
Add for each additional 6 feet in depth besides the price given, the sum, per cubic yard	10	15	20
Spreading and leveling in layers not exceeding 12 inches in depth, per cubic yard	7	10	13
Add to last item for well tamping, per cubic yard	10	10	10
Returning earth, spreading, tamping, exclusive of carting or wheeling, per cubic yard	15	15	15
Labor only, for ditto, per cubic yard	12	12	12
Paddling walls, filling cofferdams tamping clay in layers 8 or 9 inches thick, per cubic yard	..	..	$2.00
For labor only, in above	..	..	40
Clay tempered and puddled 6 or 7 inches deep, well tamped in place, per yard super	..	..	90
Covering slopes, terracing with good soil in layers about 6 inches deep, per super. yard	..	10	..
Sodding same and furnishing sod and leveling same, per super. yard	..	40	..

CARTING AWAY SUPERFLUOUS MATERIAL

Wheeling or carting stuff from excavation in addition to the foregoing items, not exceeding twenty yards distance, including filling of wheelbarrows, carts, etc., and depositing solid contents on the ground, per cubic yard..	$0.12
Add for wheeling or removing every additional 20 yards, up 100 yards from starting point, per cubic yard..........................	.06
Basketing earth or rubbish of any kind from the inside to the outside of a building, any floor, per cubic yard..........................	.15
Removing to a distance, not exceeding 1 mile, including loading carts, wagons, etc., and depositing same from vehicle, per cubic yard.	2.00
Add for every additional mile, per cubic yard....	.60
Carting away rubbish and unloading, distance not to exceed 1 mile, per cubic yard............	1.50
Add for every additional mile, per cubic yard....	.50
Loading or unloading barges, scows, or boats of any kind, alongside the stuff being delivered, within 12 yards of barge, etc., per ton......	.25
Removing by barges, scows, boats, etc., to a distance of 1 mile or under, per ton...........	.40
Add for every additional mile, or part of a mile, beyond the first..........................	.15
Cost of driver, horse, and cart, per hour........	.35
Cost of wheelbarrows, per hour...............	.01
Cost of team, wagon, and driver, per hour......	.35
Other appliances, cost must be ascertained before putting in the tender for work.	

CONCRETE WORK

Concrete should be composed of pure clean water, broken stones, or ballast or clean pit-gravel, with such a proportion of sharp sand as will fill the voids between the stones or gravel; and this latter should not be larger than such as will pass through a ring 1¾ inches in diameter. The proportion should never be less for Portland cement than one to six parts of stones and sand combined, and the concrete should be thrown into position steadily and as evenly as possible and tamped down in layers not more than twelve inches thick. The following prices include mixing, wheeling, throwing in place and tamping down. Of course something will depend on the cost of cement, and on the cost of aggregate, i.e., broken stone and sand.

CONCRETE FOR FOUNDATIONS AND PAVING

Foundations for walls, etc., circular, straight, or in thick pieces, per cubic yard	$4.60
Above foundations, underpinning, retaining walls, or similar work, per cubic yard	4.80
Blocks of such size and shape, if square, as may be required, and set in Portland cement, moulds included, per cubic yard	5.80
Foundations for paving on with brick or stone, 4 inches thick, per yard super	.65
Ditto, 6 inches thick, per yard super	.90
Ditto, 9 inches thick, per yard super	1.05
Ditto, 12 inches thick, per yard super	1.20
Floating surface of concrete and bringing it to a fair face, per yard super	.25
Add for work if executed between high and low water mark, including full protection against tides, or streams, per cubic yard	1.00
Add for every 10 feet hoisted above the level of first floor, for each cubic yard	.50

100 cubic feet of solid stone, when broken so that the largest piece will pass through a ring 1¾ inches in diameter, will equal 189 cubic feet.

Through a 2-inch ring, will equal 182 cubic feet.

Through a 2½-inch ring, will equal 170 cubic feet.

CONCRETE FLOORS AND ROOFS

The concrete for floors, pavements, roof-gardens or roofs, should be made in the proportion of one part

Portland cement, four parts of broken bricks, slag or other porous aggregate, and should be small enough to pass through a ¾-inch ring; but no sand should be used. Fine ashes from the smith's forge make the best material for this purpose, but it should not exceed in bulk one-third of the whole mass. The concrete should be laid in position gradually and continuously, until the whole work is done, and should be tamped concurrently as laid in place. Concrete under boarded floors, tile or brick pavements should be as above described, but in the proportion of one part Portland cement to five parts of aggregate, which, after being thrown in place, should be leveled off nicely and tamped down with a wooden pounder until it becomes pulpy and the "fat" or cement portion is brought to the surface, when it should be floated or finished to a fine smooth face with a wooden float.

PRICES FOR CONCRETE FLOORS AND ROOFS

Item	Price
Concrete floor, as before described, 4 inches thick, laid complete, per yard, super	$1.30
Concrete roofs, per yard super	1.00
Add for each inch in thickness above 4 inches	.18
Add if surface is finished with granite siftings, ½ inch thick	.10
Add to floors or roofs, when the under side is exposed and rendered fair with lime putty for limewhiting	.12
Concrete bed under wooden floors, ground level, as described, 4 inches thick	1.00
Chases left in floors or roofs for expansion by inserting battens, including use of same, fixing and removing, and filling up cavity with concrete, and making good surface after removing battens, per foot run	.08
Forming channels in concrete floors or roofs, not exceeding 6-inch girth, per foot run	.11

Extra to forming 4-inch projection to 6-inch flat concrete roof, and throating on under side, per foot run	.10
To these figures add for hoisting every 10 feet in height, after the first 10 feet, per yard super	.05

EXCAVATING FOR TRENCHES, DRAINAGE, FOOTINGS AND SIMILAR WORK

As before stated, the prices given in this work are not to be considered good for all time. The prices given to-day will be found quite unreliable in a month or two, or when applied to another locality. The prices, however, I do affix to the work specified may be considered moderate and fairly safe for competitive tendering, but it is always best to vary these prices by local quotations and current rates.

I have already given a few instructions to the intending contractor with reference to excavating, but it may be well, even though I may lay myself open to the crime of repeating myself, to reiterate in some measure those instructions and warnings.

The plans of the intended specifications should be well studied and specifications carefully read over, so as to thoroughly understand what the architect desires, and when things are not properly digested the architect should be consulted.

The site of the intended building should be visited, so that the nature of the soil may be known, the distance it is to be conveyed, the state of the roadway, and the distance the building materials have to be hauled. See to the levels, and ascertain as nearly as possible the amount of material to be removed. Sometimes, in digging, a very different soil reveals itself to that taken; there are sometimes loose sand, running water, rock, and other obstacles that have not been considered, and the price per yard for digging, removal of loose material, strutting sides of trenches, pumping, and cost of carting may make a considerable difference. The builder who knows the locality or site and the sub-soil is, of course, in a better position than others who tender. On some sites sand may be found a few feet from the surface, and this may be valuable and make a difference to the price; or it may be the sand has been screened and placed again on the site and covered with loam, in which case the excavations will have to go down to the "virgin" soil.

The cost of materials should be obtained before estimating. The prices of stone, bricks, sand, lime, ballast, delivered on the site, are all-important preliminaries to correct estimating. The prices of bricks, sand, lime, etc., vary very much in different localities. To take brick work, several elements are necessary before a correct price can be affixed per rod; as, for example, the price of bricks in field, the carriage to works, if by barge or rail, the cost of loading, the freight, unloading, carting from wharf to works, the price per yard of sand delivered, and of lime, and cost of labor. If there are any terra cotta or drain pipes, the cost delivered on the site should be obtained from the maker, and the same for any iron work or other special material.

As all these elements are found to vary considerably, it is only possible to obtain an approximate price. The market prices of leading items in each trade ought to be known, and for this purpose trade lists and prices are necessary. The quotation of prices for particular items is important.

More uncertainty prevails in estimating excavator's

work than in any other of the builder's trades, owing to the various kinds of soil to be removed, if the soil is carted or wheeled a long or short way, if the excavation is deeper than 6 feet (the height a man can work), if filled in, where deposited. This item is taken according to the labor involved. It may consist simply of digging and carting, as in the excavation over the site, or of digging, filling, and ramming, as in trenches for foundation. In the latter, however, both kinds of labor are required. Thus, the "digging and carting" represents that portion of the excavation which is occupied by the wall and has to be removed, and the "filling and ramming" applies to that portion of earth which is filled in and rammed against the walls. Then it is necessary to keep such items separate, as, for example, the excavation to basements and those only on the surface, as in removing the top soil and wheeling away not exceeding say 9 or 12 inches deep. In the deeper excavations in friable soils timbering is necessary, as walling and strutting the sides of trenches, etc.

In pricing items of excavation, the depth and width of trench, the nature of the soil, and the quantity of timber, if necessary, the latter measured per foot super. on each side, must be known. Digging in gravel or stiff clay costs twice as much as in loose earth. The disposal of the stuff should be made clear. Thus, the part of the trench to receive concrete may be described as "excavation and carting away, or wheeling and spreading," the portion to receive the brick work being described as "digging to trenches, part filled in and rammed, and remainder carted away." The earth may be dug and thrown out, wheeled or basketed out, or carted away to make up other ground.

Depths of 6 feet, 12 feet, or 18 feet should be kept separate.

Wall trenches in width are regulated by the spread of the footings, usually twice the thickness of wall at base, and room enough for men to work in the trench on one or both sides, usually 6 inches beyond bottom course of footings.

Pumping and bailing out water is a speculative item, and its cost can only be approximately put down. I have shown in previous pages approximately the cost of handling loam, sand, gravel and general rubbish, and the prices given these hold good in nearly all cases, but exceptional conditions must be provided for.

For large trenches and foundation work, when the earth is filled in and rammed, it is perhaps better to make a separate item, as "excavation and returning, filling and ramming," the quantity measured from outer face of brick work to side of trench by the depth of the footings, and deduct this from total excavation.

Priced bills do not help the young estimator much. To take two or three priced bills of quantities for the same building will reveal extraordinary differences, arising from various circumstances—the position and facilities of the contractor, his nearness to the work, whether he has a large plant and staff of workmen, or is a man of small capital without resources; the prices also depend on whether the estimate is prepared with the aid of drawings or specifications, or simply from a bill of quantities, from the items of a day or measured account. A man may be an expert quantity taker who has not mastered the fundamental elements of pricing; the two processes are different. The expert in prices must be a man naturally addicted to study and compare values, to analyze the composition of items; he must be able to arrive at a price by a calculation in detail. A mind so trained will be able to trace analogous conditions, will be able to generalize and compare. We should recommend the young estimator to master the contents of every trade list of materials and goods, and these should be kept, classified and indexed, on some system for easy reference. The trade and cash discounts, railway rates, cost prices, etc., should be collected and indexed for reference, and for this purpose an alphabetical index or commonplace book ought to be kept. A book for each trade should be kept to enter prices, data, and information, always giving date. Note especially the time expended on every kind of labor, as, for example, the time taken by a laborer in digging a yard cube of clay or other material, how many yards he can do in a day; the time it will take a joiner to frame a door of a certain thickness per foot super., or the time it takes to do any unit of work.

Large quantities of material, like sand or ballast or bricks, can be procured at a cheaper rate than small supplies, and a difference of at least 10 per cent in the cost may be made; but in every particuar instance it is better to make inquiries and obtain quotations from reputable dealers and contractors.

The presence of sand on the site will often save much carting away, as the sand and ballast can be used for concrete and brick work, and before pricing items of excavation inquiries should be made as to the depth of the sand below the ground level. All above the sand has to be carted away; it may be half or two-thirds of the whole depth excavated. When sand occurs in the trenches and site considerable saving is effected, and the exact quantity of this should be ascertained before pricing, so that an allowance can be made. Thus, in trenches say half full of good sand, one-half only of the quantity or of every yard would have to be carted away. The other portion will be a distinct gain. The sand should be valued at so much per yard cube, added to the saving of carting, so that there should be a great saving. It is better to provide that a certain sum shall be allowed by the contractor for every yard of sand found on site and used in the building.

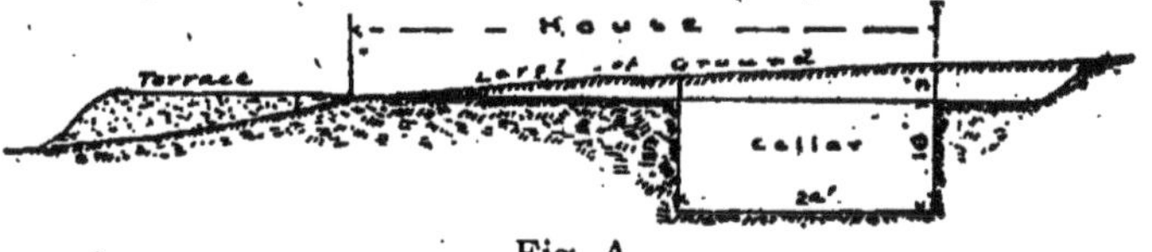

Fig. A

At Fig. A, I show a section of a site that was supposed to be irregular, and where the cellar excavation and irregular ground is shown to be removed and terraced in front of the house. This will give some idea of the proper method to figure on excavating of that kind and how the material may be disposed of.

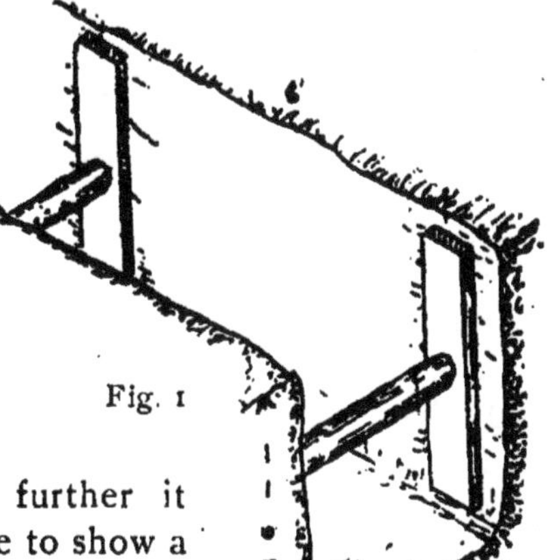

Fig. 1

Before proceeding further it may not be out of place to show a few examples of deep trenches for drainage or other purposes, cut in various sorts of ground, and the methods employed of holding the backs or sides of trench in place until the work is completed. Fig. 1 shows a trench, 3 feet 6 inches deep and 3 feet wide, that is prevented from caving in by the use of cross struts and planks placed at a distance of about 6 feet. This trench is supposed to be dug in good solid ground. These struts and planking will require about 10 feet of material for every 6 feet in length of the trench, and about one-half hour's time in putting in place and preparing stuff.

Fig. 2

Fig. 2 shows a "heading" for good ground. This, it will be noticed, is sheet-piled on top and two sides. These timbers must be sized to suit the size of cutting, and character of ground; so price must be gauged accordingly. Cost per running foot, about 65 cents.

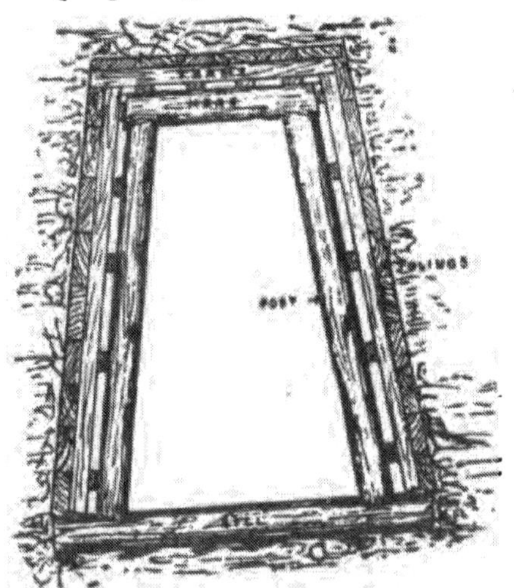

Fig. 3

Fig. 3 shows another heading. This is for very bad ground, and is supposed to be made very strong. This is an expensive affair; but the materials for use in the framework, when carefully removed, may be used again for the same or similar purposes. This style would cost about $1.50 per running foot, exclusive of digging and removing material.

Fig. 4 shows a method of shoving a ditch or trench for loose earth. This may be built with the sheet piling in two lengths, as shown. The cost of this style of shoving would be considerable and depends somewhat on the depth of cutting. Supposing this to be about 6 feet deep, the cost would be from 75 to 85 cents per running foot, which would, of course, include both sides and cost of planking and poles. Some allowance would have to be made for the return stuff, as most of the material could be used again for a similar purpose. The prices given do not include digging or removing the loose earth, but simply the shoving and the material used; but these prices will vary with the locality and cost of material and labor.

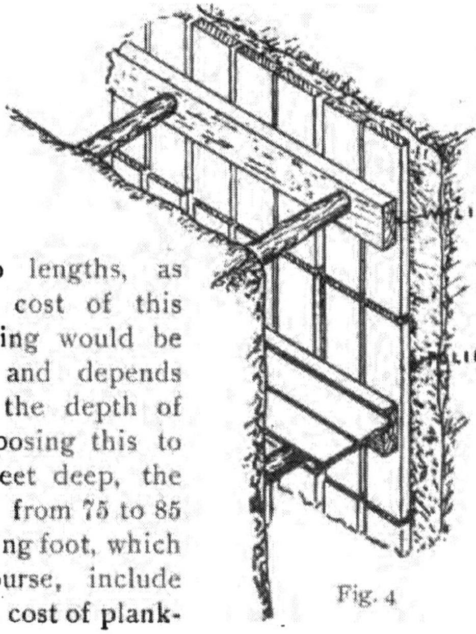

Fig. 4

The trench shown at Fig. 5 differs from those previously shown, inasmuch as this exhibits a trench with sloping or inclined sides. This is arranged for a trench dug in loose or treacherous ground, and if made about 6 feet deep the labor and materials required to complete the shoving would cost, in round numbers,

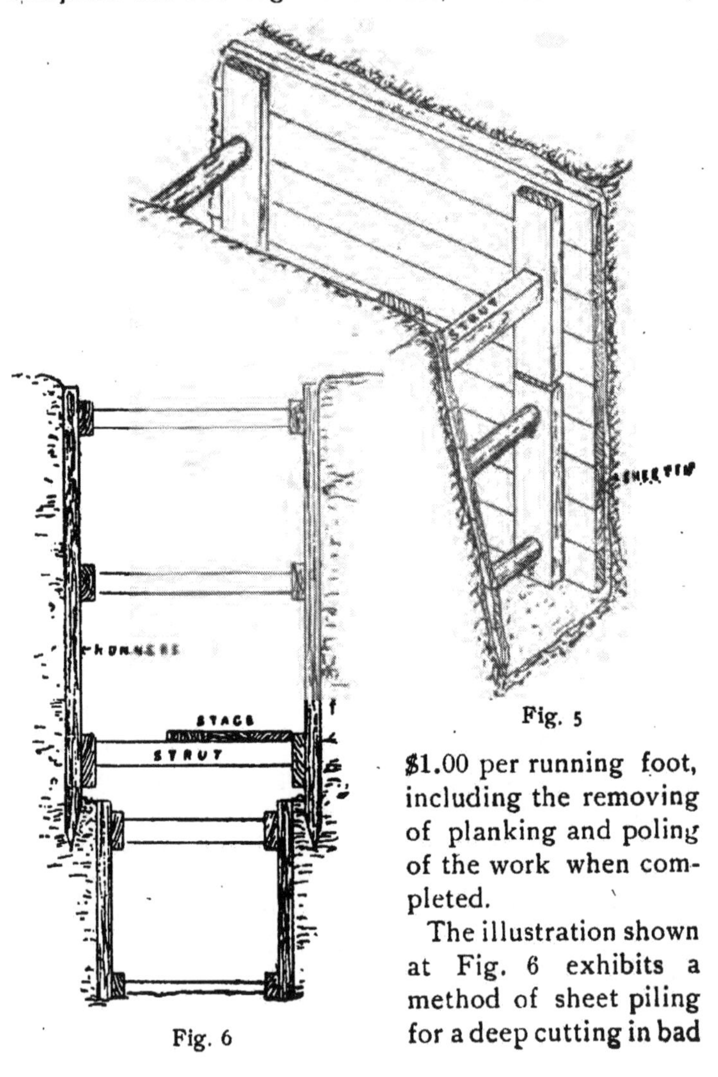

Fig. 5

Fig. 6

\$1.00 per running foot, including the removing of planking and poling of the work when completed.

The illustration shown at Fig. 6 exhibits a method of sheet piling for a deep cutting in bad or loose ground. This is an expensive necessity wherever it has to be undertaken, and requires two stories, as it were, of shoring and an extra widening of the trench at the top. The sheet piling is of plank two or three inches thick, as may be required, and each plank is pointed at one end and is driven into the soil at the bottom of the wide trench as shown, and is strutted and made secure, after which the lower trench is excavated and secured with piling and struts as shown. This style of planking and securing the work is very expensive, and each case must be figured out for itself; the cost depending largely on depth and width of trench and quality of earth to be supported. I have known of such work as described costing $6.50 per running foot for labor and materials for the purpose; the trench being about 14 feet deep on an average. This was exclusive of digging and removing the earth from the trench. Under the circumstances, it would be folly to give any stated price for this work. An approximate cost can only be obtained by actual figuring on the particular work to be done, and it is always the surest way, in cases like the one under notice, to make no allowance for returned material, for, when taken out of the trench, it will have but little value for any other purpose.

A FEW THINGS WORTH KNOWING REGARDING EXCAVATING

The following items should aid the estimator in determining prices and arranging for space, etc.

Natural slopes (with horizontal line):

Moist sand	22°	Chalk	55°
Dry sand	38°	Rubble	45°
Vegetable earth	28°	Well drained clay	45°
Shingle	39°	Wet clay	16°
Gravel	40°	Loose peat	14°
Compact earth	50°	Firm peat	45°

Height of perpendicular face which various soils will retain for a short time without falling:

Clay	9 ft. to 12 ft.
Drained loam	5 ft. to 8 ft.
Ordinary earth	2 ft. to 3 ft.
Dry sand or gravel	1 ft. to 2 ft.

In trimming banks for a permanent surface the slope should not be uniform, but flatter at the lower than the upper part. For instance, in the same soil (clay), a bank 5 feet high may stand at a slope of 1½ to 1; 10 feet high, 2 to 1; 20 feet high, 3 to 1, with practically the same permanency. The most economical section for a deep cutting or hillside would be a slope ranging from 3 to 1 at bottom to 1½ to 1 at the top.

Equivalents of slopes:

½ to 1=63° 30′	1¾ to 1=29° 44′
¾ to 1=53°	2 to 1=26° 44′
1 to 1=45°	3 to 1=18° 25′
1¼ to 1=31° 40′	4 to 1=14° 12′
1½ to 1=33° 42′	

Increased volume of earth in embankment over the same unmoved:

Sand	1/7 more.	Clay	1/8 more.
Gravel	1/11 "	Large rocks	½ "
Chalk	⅓ "		

A usual allowance for settlement is one inch for every foot of height, but the settlement is sometimes as great as 3 inches per foot.

A good excavator will dig and throw into a barrow in a day of ten hours:

In common ground	from 8 to 10 cub. yd.
In stiff clay or firm gravel	" 5 to 6 "
In hard ground (picking required)	" 3 to 5 "

In excavating, a vertical throw is taken at 6 feet, and when a trench exceeds that depth, stages must be provided. In practice, stages are usually set at somewhat closer intervals.

Clay invariably swells on exposure of the face in an excavation, and allowance must be made for this in certain works, as in well-digging and tunneling.

In calculating the quantity of excavation in a trench which tapers in depth or width, the prismoidal formula should be used, viz., area of two ends plus four times middle area, and the total multiplied by one-sixth of the length.

For an irregular site take spot levels, join all up into triangles, then multiply the mean depth of each triangle by its area.

A run is a certain distance for wheeling excavated material. With a length of one run, two barrows can be kept going without waiting. The length of a run is commonly taken to be 20 yards, but according to some it is only 18 yards, while in some districts 22 yards is allowed, and in U. S. government work 25 yards make one run. If wheeled more than three runs, a higher proportionate price has to be paid.

WEIGHTS OF MATERIALS

27 cubic feet = 1 load, and contains 27 striked bushels or 21 heaped bushels.
54 cubic feet = 1 double load.

21 cubic feet of river sand (as filled into carts)	weigh 1 ton.
22 cubic feet of pit sand (as filled into carts)	"
22 cubic feet of common ballast	"
23 cubic feet of coarse gravel	"
24 cubic feet of clean shingle	"
28 cubic feet of stiff clay	"
28 cubic feet of marl	"
29 cubic feet of chalk (in lump)	"
33 cubic feet of earth (mould)	"

A tip cart will hold about ¾ yard cube.

A wheelbarrow contains $\frac{1}{10}$ yard cube.
A small earth wagon will hold 1½ cubic yards.
A large earth wagon will hold 3 cubic yards.
1 yard cube of solid earth or gravel contains 27 striked bushels before digging, and 27 heaped bushels when dug
49 square yards = 1 rood of surface digging in country.

I have shown some of these tables in different forms in order to meet the several local customs of dealing with the same conditions; a method which, I think, will give this little work wider range than it would otherwise have.

I now offer some short rules on excavating that may sometimes be found handy:

A 10-ton locomotive steam crane excavator, fitted with a 1½-yard cube digging bucket, will excavate and deliver into wagons from 800 to 1,000 cubic yards per day of 10 working hours according to the nature of the ground.

Work in trenches costs 20 to 30 per cent more than digging over areas where the labor is not cramped. The soil is merely deposited at a safe distance (of say 2 feet) from the edge of the trench, from whence it is wheeled or carted away. Take common ground, a man would here be able to manage only 8 yards cube in one day, as there is a limited space to work in and the soil has to be pitched out one "throw." Earth that is loose enough to shovel out without using the pick, and where only one "throw" is required, may be removed for about 12 cents per yard cubic, or for less, where a plow and scraper can be employed. With the aid of plow and scraper, earth may be removed anywhere less than 100 yards for about 16 cents per cubic yard. If loaded in carts or wagons, it will cost from 20 to 30 cents per yard. Very hard clay, gravel or hardpan may cost from 40 cents to $1.00 a yard to remove. Rock will cost from $1.00 to $5.00 to remove, depending on the kind of rock. Old foundations, when stone, brick, old timber and lath, etc., are buried in mortar and other debris, will cost from 50 cents to $3.00 to remove a cubic yard from the ground to a distance not exceeding 100 yards. This includes digging, loading, chopping and unloading.

SOME ROUGH APPROXIMATE PRICES

Digging in ordinary soils, not more than 6 feet in depth, per cubic yard	$0.18
Ditto, above 6 feet in depth, and not exceeding 10 feet, per cubic yard	.21
Ditto, above 10 feet and not exceeding 14 feet in depth, per cubic yard	.25
In heavy soils, allow extra, per cubic yard	.05
Preparing for foundations, including filling in and ramming, per cubic yard	.25
Reducing the ground to the required level, the average depth not to exceed 18 inches, per yard super	.12
Wheeling ground, clay, or gravel in barrows, 20 yards run, or less	.06
Ditto, for every other 20 yards, or part of a run beyond the first 20 yards	.05
Carting and shooting, or delivering ditto, not exceeding 1 mile	.75
Ditto, for every additional mile or part of a mile	.25
(Tolls if any, to be charged.)	
Calculate wells, not exceeding 8 feet in depth, at per foot run—*i.e.*, on the depth.	
Digging and staining dry, in half a brick 4 feet 6 inches in diameter	2.00
Ditto, 5 feet 3 inches	2.40

While the foregoing on "excavating" does not cover the whole ground, sufficient has been advanced to enable the estimator to get a good idea of the requirements to make a tolerably fair estimate of the cost of any excavations that he may be called upon to figure up. As I have before stated, the thing in estimating to insure fairly correct results is "sound judgment" added to experience. The rules and methods, published in this and other work, on estimating are simply the tools with which the estimator works. If he be a good workman, a man of judgment, he will make a good job; if not, no matter how good the tools may be, the work will show up bad, and the contractor will feel himself poorer when the work is finished than before he started.

LAYING DRAIN PIPES, WEEPING TILES, ETC.

The size of drains are determined by the quantity of sewage to be conveyed and the velocity of the sewage flow. No house drain should, however, be less than 4 inches in diameter. They should be laid in perfectly straight lines with an even gradient from point to point, the necessary junctions or changes of direction being within convenient inspection chambers or manholes.

The velocity of the flow of sewage in ordinary house soil drains should be about 4½ feet per second (270 feet per minute) when flowing full, so that they may be self-cleansing when only a normal quantity of sewage is passing through them.

The quantity of sewage and waste water to be removed from dwellings, for all purposes, varies from 25 to 40 gallons per person per 24 hours. The drains should be large enough to remove one-half the estimated total daily volume of sewage within six hours.

Rainfall.—The provision for rainfall should be varied according to the district, the average annual rainfall for which can be ascertained. Rain-water drains must be sufficiently large to conveniently remove the whole of the water which may be expected to fall during the prevalence of a heavy storm.

The average rainfall from roofs in this country may be taken at 16 inches per annum, after allowing for loss by evaporation, absorption, etc.

Provision should be made for removing rainfall per hour as follows:

From roofs (measured horizontally)	75	inches in depth.
From paved surfaces	.75	"
From gravel surfaced	.4	"
From meadows or grass plots................	.1	"

For ordinary houses, drains having 4-inch branches and 6-inch mains are generally sufficient. Villas and large houses usually require larger mains, but pipes of the smallest size which may be considered adequate should be used, as being more self-cleansing than larger pipes.

An easy rule to remember for the purpose of determining the gradients of drains so as to secure good, self-cleansing velocities for the sewage, is the following well-known "decimal rule." Multiply the diameter of the pipe by 10, and the result gives the gradient for the drain, viz.:

Diameter of Drain.	Gradient of Drain.
4 inches	1 in 40
6 "	1 in 60
9 "	1 in 90
12 "	1 in 120

The maximum velocity and discharge of sewage from ordinary drain pipes (i. e., when running nearly full), as calculated by the Etyelivein formula is as follows, viz.:

Diameter of Pipe.	Fall.	Maximum Velocity per Minute.	Maximum Discharge per Minute.
Inches.	Feet.	Feet.	Gallons.
4......	1 in 40	284	146
	1 in 50	254	131
	1 in 60	232	120
6......	1 in 60	287	328
	1 in 70	265	303
	1 in 80	249	284
9......	1 in 90	284	742
	1 in 100	270	705
	1 in 110	257	670
12......	1 in 120	285	1318
	1 in 150	255	1177
	1 in 200	221	1021

Flushing.—Where self-cleansing falls cannot be obtained for the drains, periodical and, preferably, automatic flushing should be resorted to.

Rain-water drains.—Where drains are solely used for rain-water, much less fall is required than for sewage. Generally, a velocity of 2½ to 3 feet per second (150 to 180 feet per minute) is sufficient in order that the ordinary dust and dirt may be readily washed away; but the amount of water to be removed in a given time must be allowed for. The drains should be surrounded in concrete when passing through buildings or near the roots of trees, or wherever they are likely to be disturbed.

Drains should be kept as far away as possible from buildings, so that the pipes and joints may not be injured or disturbed by any settlement of the walls. By this means the risk of sewage or sewer air penetrating within the buildings is minimized. For similar reasons the drains should not pass under houses except when absolutely unavoidable, and in such circumstances heavy cast iron pipes with caulked lead joints should be used.

An unyielding bed on which to lay the drains is necessary to ensure sound and permanent work. A layer of concrete should therefore be provided under the pipes, unless the ground is naturally very hard and compact.

Branch drains should not join the main or collecting drains with level inverts. The junction should be effected within an inspection chamber or manhole, and the branch channels arranged to discharge over the channel of the main drain. Care should be taken that the branch channels are placed so that they do not discharge immediately opposite each other when entering the main channel.

Stable drainage should be kept separate from the house drainage in all cases where practicable.

Covers to inspection chambers should have a clear opening of 24 by 18 inches, so that a man may conveniently pass through them.

The materials of which drain-pipes are made varies considerably in different localities. Well-burnt stoneware pipes of good quality are thoroughly vitrified, and when broken present a fine close grain with a somewhat metallic appearance. Fire-clay pipes do not possess such a dense and close grain, and are more absorbent than stoneware pipes. Earthenware pipes are quite unsuited for use in house drainage.

Stoneware drain-pipes should be of the description known as "salt-glazed," so as to obtain an impervious and lasting surface. For ordinary house drainage purposes the pipes are usually made in 2-foot lengths.

Specially selected and tested stoneware pipes in 3-foot lengths may be obtained from manufacturers at a slight additional cost over ordinary pipes. "Tested" pipes should be capable of withstanding a pressure of 25 feet head of water without showing signs of sweating.

Neat Portland cement is generally used for jointing ordinary spigot and socket pipes, or cement and sand in the proportions of one part cement to one part sand.

Cement joints must be very carefully formed and wiped out as the work proceeds, so as to avoid burrs on the inside of pipes.

Greater security is obtained by adopting one of the several well-known forms of patent safety joints now made by the leading manufacturers. They are more expensive than pipes with ordinary spigot and socket joints, but the advantage of obtaining a stronger and safer connection more than counterbalances the additional cost.

Protection against fracture can best be obtained by entirely surrounding the pipes with concrete. A thickness of 6 inches of concrete is usually sufficient for this purpose.

The average thickness and weight of glazed stoneware drain-pipes per 2-foot length is as follows, viz.:

Diameter of Pipe.	Length of Socket.	Thickness of Stoneware.	Average Weight per 2 ft. Length of Pipe.
Inches.	Inches.	Inch.	Lbs.
4	1½	½	18
6	1¾	⅝	32
9	2	¾	58
12	2	1	90

The cost per foot of these pipes should be obtained

from the dealer, along with the extra cost of Wys, V's or other connections that may be required, before any estimate is made. If the drain-pipes are to be laid in concrete, the cost of the concrete and labor of putting it in place must also be added. The digging of trenches has been dealt with before, but in making an estimate this item of digging and removing the soil must not be overlooked. It is not possible to give a price for work of this kind unless the size of pipes, depth of trench, if or if not bedded in cement or concrete, etc., are given; then a price per foot in length may be arrived at.

Cast iron pipes are largely used in high-class drainage work. The cost is not much more than that of good glazed stoneware surrounded with 6 inches of concrete.

The advantages obtained by the use of cast iron pipes as compared with glazed stoneware are as follows:

1. The pipes are of greater strength. They are consequently not so liable to become fractured or broken.

2. Air and water-tight joints can be readily made by running with molten lead and caulking.

3. Fewer joints are required, owing to the longer lengths of the pipe.

For substantial work the iron pipes should be of similar thickness and strength as those used for ordinary water mains. They are generally laid in 9-foot lengths, with spigot and socket joints run with lead and caulked.

Whenever a drain passes under or through a wall it should be of iron, then if any settlement takes place the iron will offer a much greater resistance to the consequent pressure than glazed earthen tiles would.

Weeping tiles may be common field tiles, or they may be ordinary drain tiles of small diameter. They are made use of occasionally to drain around a foundation wall, or to drain under the concrete floors of a cellar. When field tiles are employed they butt at the joints, which are not made tight, as water is intended to enter the pipes at every joint. The same, also, with ordinary tiles, the joints being left loose so that water may enter at every joint.

The cost of laying weeping tiles is very small, as a man will lay 30 or 40 feet per hour, but the cost of the tile themselves must be considered. There will be no excavating for these tiles, as, in the case of a cellar, the tiles are laid on about the same level as the foundation; the tiles are laid on a level, and against the footings. Of course, the tiles in both cases must lead into the main drain, and this may necessitate some extra digging.

FOUNDATION FOOTINGS

In placing footings a special rate should be made, as much more care and time is required in getting good flat stones of the proper thickness, and leveling them on their beds, than in laying an ordinary wall. In my own practice I have usually charged up 50 per cent more per cord for footings than for the other portion of stone wall, and this additional charge has been found not a bit too much in most cases. If the footings are of concrete, as is generally the case now, then this must be charged in accordance with the rules given under the head of concrete. Concrete footings may be flat or they may have a broad base and narrow top, just wide enough to take the walls, whether of brick or stone.

The three illustrations shown at Fig. 7 give an idea of both concrete and stone footings. The first is con-

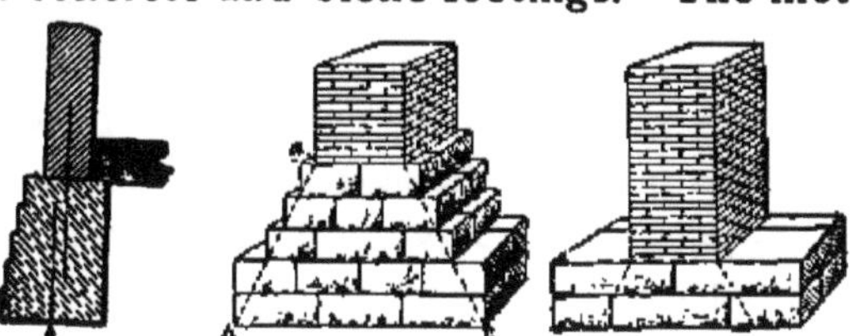

Fig. 7

crete having a rectangular section, or it may be inclined from the outside. The second is formed of five thicknesses of dimension stones drawn in towards the top. This is intended to carry a very heavy wall. The third is formed of two thicknesses of dimension stones, but is not drawn in. All three of these are good examples for footings, but they do not by any means cover the whole ground; another example is shown at Fig. 8. This is a section and is intended to carry a high and heavy wall. The concrete is 18 inches thick and fully 5 feet 6 inches across. In estimating for this, the concrete must be figured at so much per cubic yard, and full allowance made for wheeling and dumping. The brick or stone work above, until the level of the ground is reached, should be charged up about 10 per cent above the regular rates.

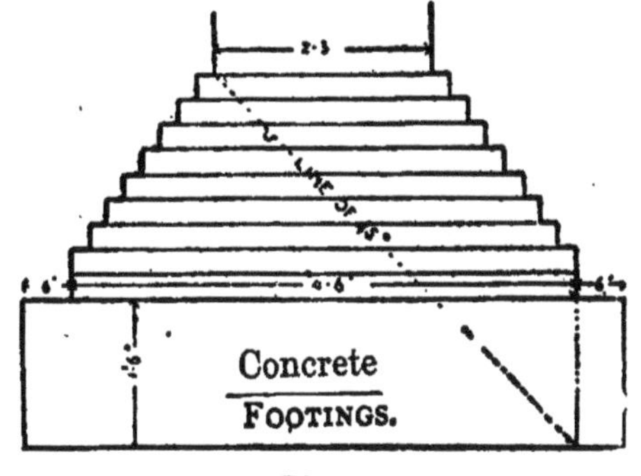

Fig. 8

If footings are laid in with ordinary quarried stones

without dressing, the cost will be about $5.00 a cord of 100 cubic feet, exclusive of all materials. Cost of materials, stone and mortar, to be added, and if laid in cement the cost will be a trifle more.

Ordinary quarried stone laid in the wall, set in good lime mortar, is worth at the present time $5.00 per cord of 100 feet in the wall for setting.

In buying rough quarried stones by the cord, which is the usual way in most of the states and Canada, the purchaser is supposed to receive 128 feet cubic in the rough, but the mason's measurement, including mortar joints, is 100 cubic feet in the wall for one cord; and when he buys he expects to pay for the 100 feet per cord and to receive pay on the same basis.

In putting in stone foundations as above, the estimator must make provision for all openings, and when cut stone or cement sills and lintels are used for doors or windows, they must be charged up extra by the running foot. All ventilators must be extra items and duly charged. Figure for all openings for drain-pipes, water, gas, or other pipes entering the basement or cellar. All areas must be figured on by the yard super., if in cement or stone, according to prices given; steps, walls and copings must all be measured off and charged up according to size and material. Prices, if not found in this work, must be ascertained in the locality where the work is to be executed.

Sills and lintels, in either stone or cement, may be bought from the dealer by the foot super. or cubic foot, and price lists of same may be obtained from the manufacturers.

Footings and basement or cellar walls are sometimes specified to be made damp-proof, and the architect *sometimes* shows how the walls are to be constructured so as to be damp-proof. I show two methods, both of which are expensive but certain in result. Fig. 9 shows a concrete footing with a section of concrete carried up the walls to the height of top of cellar floor, which is also of concrete 4 or 6 inches thick. A damp-proof course of slate or asphalt is shown on a line with cellar floor, and is continued on the outside wall to a point above the line of ground. This is an effective method. In this case the concrete is worth from 10 to 15 per cent more to put in place than if a simple footing as above. Damp-proof course is worth from 15 to 25 cents a running foot, according to the thickness of the wall.

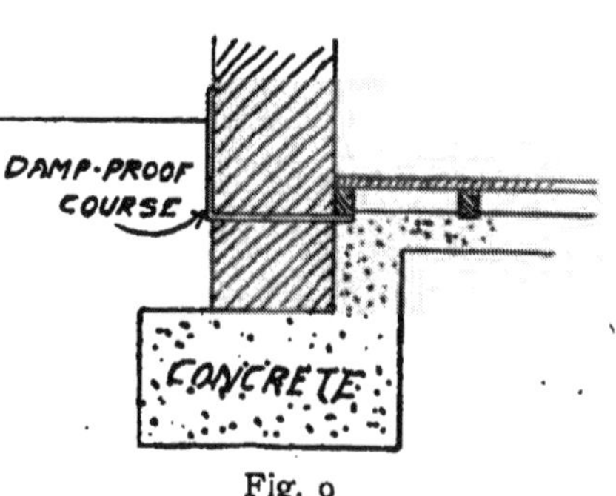

Fig. 9

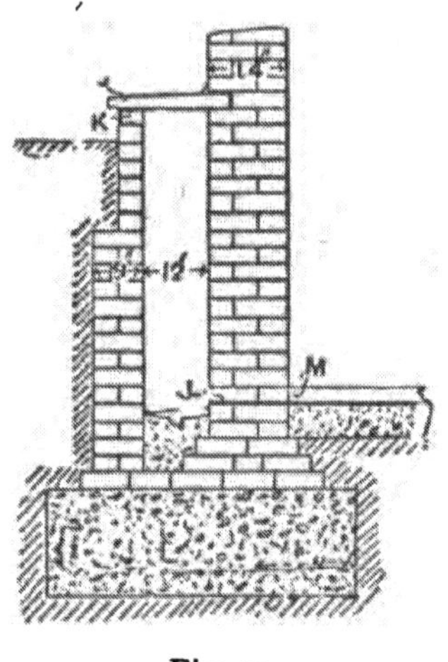

Fig. 10

The footing shown at Fig. 10 is a still more expensive one. Here is a wide footing in concrete and a double wall for a portion of the height. There is a damp course of slate laid at L in the main wall and level with the finishing coat of cement, M, on the cellar floor. The outside wall, R, is simply to hold back the soil on the outside, thus forming a 12-inch space between the walls for air and to avoid damp. Concrete is thrown in between the walls below to a thickness of 8 or 10 inches, thus preventing any damp from attacking the main wall. The space between the two walls is covered over at the top with a stone slab, J, which prevents any rain or other water from gaining entrance.

Here we have several new items to figure on. Extra bricks in outside wall, covering slab, concrete in air space and damp course. Figure the concrete by the yard cubic, the slab by the running foot, the extra bricks in the usual manner and the damp course same as before mentioned.

We are now in a position to describe some of the methods of estimating as mentioned in previous pages, and will endeavor to do so before entering into detail estimating.

As I have stated before, there are five distinct methods of estimating, namely, by rough quantities, by the square, by the unit of accommodation, by cubing, and by itemizing details. The two latter may be considered the best methods of the five, and the last the best of all, though the most troublesome. Of the first three I will say but little, as they will be apt to lead the ordinary contractor into a maze of difficulties that will eventuate in loss of time and money; besides, a fairly correct description of them and the method of using them have been already presented. It may be well, however, to make a few remarks concerning them.

The cost of buildings is constantly changing, so it must be remembered that no matter what prices are given in this book the estimator must in every case use his own judgment and true knowledge when making up his tender, and add or deduct whatever percentage may be necessary to suit the fluctuations in prices of labor and materials. During the last decade the cost of buildings of every kind has increased from 30 to 40 per cent; stone and the more elaborate buildings have increased in a greater proportion than the cheaper kind, owing perhaps to the greater cost of expert labor and the more luxurious fitments. Some idea of the cost of a proposed building may be derived from a study of the proportional cost of the various trades. Of course the result will only be approximate. For instance, in ordinary domestic buildings the brick work and masonry will represent from one-third to one-half of the total cost, unless the building is a frame one, in which case the wood work, including labor and hardware, will represent about three-fifths of the total cost. The following figures show, from actual experience, about the average ratio of costs of the various trades for the erection of brick or stone dwellings with slate roofs.

	Percentage of total cost.
Excavator and drainage	3.0
Bricklayer and mason	36.0
Slates and roofs	4.5
Carpenter, hardware, etc.	34.0
Electric wiring, bells, and fitments	4.0
Plasterer, stucco work, etc.	6.0
Plumber, heating, etc.	8.0
Painter, glazier, paper hanger	4.5
Total	100.0

Similar tables may be constructed showing the average ratio of cost for each of the trades in the erection of public buildings, schools, churches, theaters, etc., and these tables will prove of great assistance to the estimator when figuring up for buildings of a similar nature. Having the total cost of one building of this kind, with the cost of each of the trades named, on the same building, the rest is easy, the difference in size and character of the two buildings being considered. This may be considered estimating by comparison. If the brick and stone work of one building costs, say 10 cents per every cubic foot of the building, then the wood work will cost, according to the rule given, about 8 cents and a fraction for a foot cubic of the whole building, and the other trades in proportion as laid down.

This method is rather arbitrary, and, while given here, is not supposed to be quite correct, but when properly understood will be found quite useful.

On the same lines I give another, which may sometimes be employed in determining the cost of labor where all materials are furnished. This is a rough and ready means of making a comparison, but is pretty nearly correct and may sometimes be used to advantage:

	Percentage to total cost
Excavator	
Drainage, etc.	
Bricklayer	
Mason	
Slater or roofer	
Tiler	
Carpenter	
Joiner and hardware	
Plasterer	
Plumber	
Painter	

Here, then, by this rule we find that if the material for the painter cost one-third of any given amount,

the labor will cost two-thirds of the same amount to do the work; and the same method may be applied to other trades. The figures must be filled in to suit the current prices.

The average wages paid, at this writing, March, 1904, according to E. M. Craig, Secretary of the Building Contractors' Council, Chicago, Ill., in 29 leading cities in the United States, are given in the foregoing table. The rates given are in cents per hour, with a few exceptions, which are given in days of nine hours each.

This table will aid materially in determining the cost of work in and about the cities named.

In estimating by "rough quantities," the amount of materials and workmanship are first ascertained from the drawings and specifications in a broad and comprehensive manner, the work being concentrated as much as possible, and the whole dealt with as shown in the previous paragraphs as this method, which see.

Estimating by the square has been discussed before, but it may be briefly referred to again, as this method is quite common in some localities. This method is recommended by some authorities as being superior to cubing, as it gives a better idea as to the character of work and quality of materials, though, I must confess, I do not see where the advantages come in, for the expert "cuber" must take both those conditions into consideration when deciding on his "constant" for the cost per foot cubic of the building being estimated upon. In addition to what has been said on this method, I add the following: "The mode is to take the constructional shell only, pricing it at so much per square; walls, for instance, are taken according to their thickness and manner of finishing, whether they be wood, brick or stone. This must include all excavating, concreting, plastering, painting and paperhanging. The floors must include all joists, bridging, ceilings and ornamentations of all kinds. The roofs include all that is required to complete them, as shown on plans and described in specifications, and are measured on the slope of the rafter; and all the other work, partitions, stairways, and everything in the building, must be treated in like manner, and all reduced to squares of 100 feet super. A price is then placed on each, and the whole added together. Such a system of superficial measurement certainly has some advantages, and should be fairly satisfactory, as it takes into account the materials and labor in a fairly exact manner and form. Of course, as before stated, a special list of prices must be compiled for each set of squares, the outside walls having one price per square, the floors another, and so on until the whole of the surfaces have been priced per square. It will be seen that care and discrimination are requisite for estimating by this method, or serious errors will occur.

As an example of this method of estimating I submit the following, which is for a balloon frame building put up in the usual manner, and for convenience a space in the building is taken for a basis of 20 x 20 feet, making four squares. This basis may be taken for any portion of the work, i. e., walls, roofs, floors, etc. The studding employed is 2 x 4 inch, sized on one side and one edge. The studding is placed 16 inches from centers and covered with dressed and matched boarding. Building paper is next laid on, and then first or second clear siding is used. Plates are included in the cost and are put on double thickness.

ANALYSIS OF OUTSIDE WALLS

19 pieces, 2 x 4 inch, 20 feet long—247 feet, at $20 per M	$ 4.94
466 feet dressed and matched fencing, at $25.00	11.65
475 feet siding, at $30	14.25
11 pounds nails	.50
30 pounds paper, at 2½ cents per pound	.75
Framing and putting in place 247 feet of scantling, at $8 per M	1.98
Laying 4 squares of flooring, at 50 cents per square	2.00
Laying 4 squares of siding, at $1.12½ per square	4.50
Laying 4 squares, at 12½ cents per square	.50
	$41.07

Dividing this sum by 4 gives the price of a single square, $10.27.

The analysis of cost of four squares of roofing, the rafters being 2 × 4 inch scantling, set 2 feet between centers, covered with dressed and matched fencing, and the best quality of cedar shingles laid 4½ inches to the weather, is as follows:

ANALYSIS OF ROOF WORK

12 scantlings, 2 x 4, 20 feet long—156 feet, at $20 per M.	$ 3.12
466 matched (feet) boarding, at $25 per M	11.65
3⅓ M. shingles, at $3 per M	10.00
10 pounds nails, 3d	.63
14 pounds nails, 8d and 10d	.30
Framing and putting in place 156 feet 2 x 4 scantling, at $8 per M.	1.25
4 squares of roof boarding, at 50 cents per square	2.00
4 squares of shingling, at $1.25 per square	5.00
Staging	.63
	$34.58

This sum in turn, divided by 4, gives as the cost of a shingle square, $8.64½.

The following is an analysis of cost of four squares of flooring, laid on joists 2 × 8 inches, the flooring being selected from No. 1 fencing, and the joists being placed 16 inches between centers. Allowance is made for doubling where necessary.

ANALYSIS OF FLOORING

17 joists, 2 x 8 inches, 20 feet long—459 feet, at $20 per M	$ 9.18
466 feet of flooring, at $30 per M	13.98
15 feet of 1 x 2 inch bridging, at 2 cents	.30
10 pounds of 8d common nails	.30
3 pounds spikes	.08
Laying 4 squares of flooring, at 50 cents per square	2.00
Framing 459 feet of joists, at $5 per square	2.30
Bridging	.50
	$28.64

Dividing this amount by 4, as in the previous cases, gives $7.18½ as the cost of one square of flooring. It may be remarked in this connection that these figures are based upon present prices in Chicago.

The following is an analysis of the cost of an inside door, 2 feet 8 inches by 6 feet 10 inches, 1⅜ inches thick, cased and finished complete except the one item of painting:

ANALYSIS OF COST OF DOOR

Frame, 2-set casing and stops	$2.00
18 feet of molding, 2½ inches	.28
1 threshold, hardwood	.15
1 first quality door, size as given above	1.95
3½-inch mortised lock, bronze face, bolts and striking plate	.63
Porcelain knobs, plated roses and escutcheons	.40
1 pair of 3½ japan butts and screws	.25
Setting frame	.25
Casting up, 2 sides	.40
Putting down threshold	.15
Molding 1 side	.20
Fitting, hanging and trimming door	.75
	$7.41

The following is an analysis of cost of a four-light window, with sash 14 × 30 inches, 1⅜ inches thick, check-rail, the window set, cased and finished complete:

ANALYSIS OF COST OF WINDOW

Window frame prepared for weights	$2.15
Sash glazed	2.10
20 feet 2½-inch molding	.30
25 feet inside case and window sill	.75
28 pounds of sash weights	.56
Sash cords	.18
Grounds for plastering and putting on	.30
Setting frame	.25
Casting up	.55
Fitting sash	.15
Nails	.10
Sash locks	.25
Putting on sash locks	.10
	$7.74

This example gives the key to the method of estimating by the square, also how to estimate the cost of a door or window in place.

The prices given may not be correct for any other place but Chicago, and even then the prices may differ in each ward; so the estimator must in this, as in all other cases, be sure of his prices before closing his tender. I have known the prices for door and window frames vary as much as 30 per cent in factories not a quarter of a mile apart.

Later on I will give other examples of estimating by analysis.

ESTIMATING BY UNITS OF ACCOMMODATION

This method of estimating does very well for certain descriptions of buildings, such as churches, schools, prisons, hospitals, asylums, stables, and buildings of a similar kind, but apart from these it has no value, and its value in the cases mentioned is not by any means a fixed factor. The system is based on the known cost of buildings which give so much space to each scholar, patient, sitting, horse, or prisoner. Thus, if we know how much a stable costs that was built to accommodate 20 horses, it is a simple matter to estimate how much it cost per one horse space; for if the building complete cost $4,000, that would give the cost per horse at $200.00. So, also, with schools. If we know of a school for 100 children that cost $10,000, we know that each sitting cost $100.00; therefore it is reasonable to suppose that other schools, everything being equal, will cost $100 per sitting. It must not be forgotten, however, that conditions are not always the same, and while a "jumped" figure of this sort may be, and is approximate, it is not always correct, for no two buildings, even though they are side by side and built concurrently, can possibly be built at the same actual cost. I have seen the attempt made on several occasions, and the variations amounted to from 3 to 7½ per cent; quite a large amount if taken from the 10 per cent profits of the work.

On occasions when time will not admit of even a sketch of the proposal being made, this method affords oftentimes the only ready means of ascertaining the approximate cost. Similarly, for certain minor accessories, when the cost of materials and construction varies but slightly for units of the same class, as in a range of latrines, etc., the approximate cost can be easily determined in this manner. In order to give the reader some basis to work on, I submit a few examples of price for units, which are as near as possible average ones for the whole of the

United States and Canada, and while they may not be correct, they may be depended upon as being approximate.

Cost of each room in tenements	from	$350	to	$450
Cost of each room in cottages	"	290	"	.360
Cost of each room in residences	"	320	"	420
Cost of each room in villas, etc	"	450	"	700
Cost per patient in asylums	"	1,400	"	1,650
Cost for each soldier in barracks	"	750	"	900
Cost of churches, plain, per sitting	"	45	"	60
Cost of churches, ornamental, per sitting	"	68	"	134
Cost of first-class stables, per cow	"	175	"	195
Cost of first-class stables, per horse	"	200	"	225
Cost of second-class stables, per cow	"	120	"	135
Cost of second-class stables, per horse	"	150	"	165
Cost of third-class stables, per cow	"	75	"	95
Cost of hospitals, complete, including all offices, buildings, etc., per bed	"	1,500	"	2,200
Cost of cottage hospitals, per bed	"	1,000	"	1,200
Cost of general hospitals, per bed	"	500	"	750
Cost of isolated hospitals, including all necessary offices, buildings, and other conveniences, per bed	"	1,750	"	2,250
Cost of buildings put up in a hurry for temporary occupation, per head	"	90	"	100
Cost of latrines for barracks, per seat	"	75	"	100
Cost of city and town lodging houses, per bed	"	275	"	375
Cost of music halls for cities and towns, per head	"	75	"	125
Cost of music halls for small towns, per head	"	35	"	75
Cost of schools, complete, large cities, per scholar	"	60	"	100
Cost of schools in small towns and villages, per scholar	"	42	"	62
Cost of schools in country places, per scholar	"	35	"	45
Schools, infant schools, per scholar	"	25	"	35

Cost of theaters, complete, large cities, per seat .... from	**$80 to $110**
Cost of theaters, small cities and towns, per seat .... "	**48 " 82**

These examples are given for brick buildings of good style. If the buildings are of stone, from 10 to 20 per cent must be added, according to the quality of the stone and amount of ornamentation. There are theaters in New York, Chicago, Philadelphia, and other large cities, that cost per seat 50 per cent more than I have placed in the foregoing list, but these are exceptions to the rule.

If the buildings are of wood, that is, frame buildings, then a deduction of from 10 to 15 per cent may be made from the figures given, which will make the figures approximately correct. Theaters or other buildings, built of bricks and stone, or of bricks, stone or terra cotta, cost more than buildings built exclusively of bricks, and provisions must be made for extra cost whenever this condition exists, and much is necessarily left to the judgment of the estimator in determining the extra assessment.

ESTIMATING BY CUBING

This method, while far from being exact, is, in my opinion, a more correct method than either of the others presented. At the same time the expert estimator will frequently change his constants to suit varying conditions.

The following list of the cost per cubic foot, of buildings named, which was prepared by Mr. Kidder several years ago, and published in *The American Architect*, may be of some assistance to those who desire to know the cost of similar buildings. I may say, however, that it would be safe to add at this time at least 10 per cent on the bulk, as the prices of labor and material have advanced sufficiently to warrant that addition during the past five years.

I have added to Mr. Kidder's list a few others, but as I have been unable to get the most prominent buildings that have been erected within the last few years, this table is not complete up to date, so far, at least, as the cost per cubic foot of the more recent buildings is not included.

TABLE SHOWING DATE OF CONSTRUCTION AND COST PER CUBIC FOOT

Date		Cubic contents	Cost per cubic foot
1879	Central Music Hall, Randolph and State Sts.	1,248,000	14.4 cts.
1881	Borden Block, Offices, Randolph and Dearborn Sts.	840,000	14.9 "
1881	Brunswick & Balke Fact'y, Superior St	1,219,200	5.4 "
1882	Brunswick & Balke Fact'y, Huron St.	565,000	6.2 "
1882	L. Rosenfeld, Stores and Flats, Washington and Halsted Sts.	885,456	10.7 "
1882	Hammond Library, Ashland Ave.	183,300	19.0 "
1883	Wright & Lawther, Oil-mill, Polk St.	520,000	6.8 "
1883	R. Knisely, Stores and Flats, MadisonSt	138,320	11.2 "
1884	A. Knisely, Factory, West Monroe St.	1,412,640	6.1 "
1884	J. W. Scoville, Factory, Desplaines St.	697,000	6.4 "
1885	Zion Temple, Synagogue, Ogden Ave.	478,400	7.9 "
1887	Auditorium Building, Congress St.	9,128,744	36.0 "
1887	Standard Club-house, Michigan Ave.	916,917	12.9 "
1888	A.Loeb & Bro.,Warehouse,Michigan St.	123,300	12.9 "
1889	Jewish Training School, Judd St.	447,854	10.0 "
1891	A. Loeb & Bro., Flats, Randolph and Elizabeth Sts.	499,531	10.4 "
1891	Meyer Building, Store, Franklin and Van Buren Sts.	2,099,700	9.6 "
1891	J. W. Oakley, Warehouse, La Salle and Michigan Sts.	1,390,313	6.9 "
1891	Schiller Building,[1] Randolph St.	2,433,440	30.8 "
1893	Stock Exchange Building,[2] La Salle and Washington Sts.	3,493,500	33.2 "

NOTE: [1] Sometimes called the German Theatre, 17 stories, skeleton construction, faced with terra-cotta. Rich marble work. Theatre occupies about 4 stories. Offices above. [2] 13 stories, flat roof, skeleton construction, rich terra-cotta facing.

		Cost per cubic foot
1886	The Rookery Building, Chicago, Ill., Burnham & Root, Architects	32 cts.
	Monadnock Building, Chicago, Burnham & Root and Hollabird & Roche, Architects	42½ "
	Rialto Building, Chicago, Burnham & Root, Architects	27 "
	Masonic Temple, Chicago, Burnham & Root, Architects	58 "
	Chamber of Commerce Building, Boston, Mass	29 "
	New England Life Insurance Building, Boston, Mass	60 "
	The Hemmenway Building, Boston, Mass	43 "
	Ten Story Office Building, New York City	60 "
	Board of Trade Building, Montreal	20 "
	Ten Story Office Building, New York City	50 "
	Seven Story Office Building, New York City	37 "
	Six Story Office Building, New York City	26 "
	A similar building, one front	24 "
	Two Four Story Office Buildings, one front, New York City	47 "
	Herald Building, New York City	46 "
	Chamber of Commerce, Cincinnati	26 "
	Wainwright Building, St. Louis, Mo	24¾ "
	Union Trust Building, St. Louis, Mo	27¾ "
	Equitable Life Insurance Building, Denver, Colo	42 "
	Ernst & Cramer Building, Denver, Colo	17 "
	Masonic Temple, Denver, Colo	19 "
	Crocker Building, San Francisco, Cal	63 "
	Endicott Building, St. Paul, Minn	29 "
	Four Story Office Building, Rhode Island	38 "
	Three Story Office Building, Connecticut	50 "
	Three Story Block, Denver, Colo	8½ "
	Fourteen Story Hotel, New York City	44 "
	Brown-Palace Hotel, Denver, Colo	30 "
	Denver Athletic Club Building, Denver, Colo.	18 "
	Denver Club Building, Denver, Colo	24 "
	Public Library, New London, Conn	36½ "
	Howard Memorial Library, New Orleans	44 "
	Public Library, Toronto, Ont	22 "

Fire-Proof Hospital Building, New York....	40 cts.
Six Story Hospital Building, New York.....	32 "
Hill Theological Seminary, St. Paul, Minn...	11 "
Wingate Hall, State College, Owno, Me.....	10 "
Grammar School Building, Denver, Colo.....	9½ "
Grace M. E. Church, Cambridgeport, Mass...	8¾ "
Christ M. E. Church, Denver, Colo.........	20 "
City Dwellings (of brick) in Chicago.....17 to	20 "
City Dwellings (of wood), Eastern towns....	11 "
First-class Stone Homes in Denver, Colo....	27 "
Brick Houses, Modern Improvements.......	14 "
Cheap Brick Houses, 8-roomed, about......	10 "
Cheap Wooden Houses, 8-roomed, about....	7½ "
"Veneered" Houses, Two-story............	8 "
Rough-cast Cottages, First Class...........	6¾ "
Rough-cast Cottages, Second Class.........	5¾ "
Rough Wooden Sheds, Barns, Stables, etc. 3½ to	5 "

From the foregoing table the average cost of buildings of any description may be approximately determined. The highest figures shown are those for the Crocker building of San Francisco, Cal., the cost per cubic foot being 63 cents; the lowest amounts given being for rough wooden sheds, barns, etc., which are put down at from 3½ to 5 cents per cubic foot. These last figures seem a little large for the kind of work mentioned, but they are handed me by a builder who has had a large experience in these kinds of buildings.

While the foregoing deals altogether with the cubic foot, the same principle may be applied to yards or perches or any other fixed dimensions, and as an example I give herewith a table of miscellaneous matters that will be found very useful when estimating:

TABLE SHOWING PRICES OF WORK OF VARIOUS KINDS

Spruce lumber per M. in place on roof or floor	$25.00
H. P. per M. matched, nailed and finished on roof or floor...................	35.00

H. P. per M. matched rafters and joists finished on roof or floor				$30.00
Slate roof, no boarding, per square	from	$7.25	to	12.50
Slag and gravel roof, no boarding	"	5.00	"	7.00
Composition roof, no boarding, per square	"	2.00	"	5.00
Wood shingle roof, no boarding, per square	"	3.25	"	5.20
Tin roof, with boards, per square	"	9.75	"	13.00
Corrugated iron roof, no boarding, per square	"	7.20	"	10.00
Steel stamped shingles, no boarding, per square	"	4 50	"	6.00
Common brick work, per cubic foot	"	.28	"	.38
Public masonry, per cubic yard	"	4.00	"	7.50
Concrete, per cubic yard	"	5.50	"	8 00
Cut stone pier caps, per cubic foot	"	1 75	"	2.25
Piles driven in place, per lin. foot	"	.25	"	.30
Earth excavation, per cubic yard	"	.50	"	.52
Steel truss and column frame in place				4$\frac{1}{2}$c. per lb
Steel beams in place and secured in place				3$\frac{1}{8}$c. per lb.
Plain castings in Sit				2$\frac{1}{4}$c. per lb.
Corrugated iron No. 22 gauge, in place, per super foot				.07$\frac{1}{2}$
Galvanized iron flashings, per square foot				.14
Door frame and doors, finished, per square foot				.52
Window frames and windows, per square foot				.54
Sash, glazed and painted, per square foot	from	$0.16	to	$0.28
Gutter and conductor pipes, per lin. foot	"	.25	"	.30
Wood stairs, 3 feet wide, straight, per step	"	3.00	"	3.25
Iron stairs, 3 feet wide, straight, per step	"	7.00	"	10.00
Steel shutters, rolling, per square foot	"	.50	"	.55
Louvres, fixed, per square foot	"	.45	"	.55
Louvres, movable, per square foot	"	.70	"	.80
Sheet iron doors and shutters, per square foot	"	.35	"	.45
Skylights, $\frac{1}{4}$-inch glass, per square ft.	"	.20	"	.30

Skylights, white glass, per square ft.	from	$0.18	to	$0.20
Pipe railings, per foot in length.....	"	.45	"	.55
Ventilators, round, per foot in length.	"	4.50	"	10.50
Metal cornice, per lineal foot.......	"	.12	"	.30

It may be useful to my readers to know in a general way the cost per cubic foot of a few buildings other than those already given, and to this end the following are presented:

Public abattoirs, brick, per cubic foot,	from	$0.14	to	$0.16
Small cottages, brick, per cubic foot.	"	.13	"	.17
Country court houses, brick, per cubic foot........................	"	.22	"	.30
Lunatic asylums, including wards, etc., per cubic foot................	"	.16	"	.25
Farm barns, wood, per cubic foot...	"	.04	"	.06
Farm barns, brick, per cubic foot...	"	.07	"	.08
Armories, wood, per cubic foot.......	"	.09	"	.11
Armories, brick, per cubic foot.....	"	.11	"	.14
Armories, stone, per cubic foot.....	"	.18	"	.26
Public baths, complete, wood, per cubic foot	"	.14	"	.17
Public baths, complete, brick, per cubic foot.....................	"	.16	"	.20
Public billiard rooms, wood, per cubic foot........................	"	.16	"	.20
Public billiard rooms, brick, per cubic foot........................	"	.19	"	.24
Breweries, including all necessary machinery, tubs, cellarage, coppers, cooler, pumps, etc.—				
Wood, per cubic foot......	"	.12	"	.16
Brick, per cubic foot	"	.14	"	.18
Stone, per cubic foot......	"	.15	"	.19
Single span bridges, brick or stone, per foot super................	"	5.00	"	15.00
Double or more spans, brick or stone, per foot super................	"	15.00	"	30.00
If in granite, per foot super	"	32.00	"	50.00

Bungalows and summer cottages, wood per cubic foot	from	$0.12	to	$0.16
Bungalows and summer cottages, brick, per cubic foot	"	.17	"	.19
Plain country churches, wood, per cubic foot	"	.09	"	.12
Plain country churches, brick, per cubic foot	"	.12	"	.15
Plain country churches, stone, per cubic foot	"	.14	"	.17
Churches for cities, stone, per cubic foot	"	.21	"	.40
Coach houses, brick, per cubic foot	"	.10	"	.12
Colleges, first class, complete, brick, per cubic foot	"	.20	"	.28
Colleges, first class, complete, stone, per cubic foot	"	.25	"	.35
Colleges, second class, complete, brick, per cubic foot	"	.18	"	.22
Underground conveniences, complete, per cubic foot	"	.80	"	1.40
Stable for cows, wood, per cubic foot.	"	.08	"	.12
Stable for cows, brick, per cubic foot.	"	.13	"	.15
Stable for horses, wood, per cubic foot.	"	.10	"	.13
Stable for horses, brick, per cubic foot.	"	.14	"	.17
Power plant station, brick, per cubic foot	"	.14	"	.18
Fire engine house, brick, per cubic ft.	"	.14	"	.17
Residential flats, brick, per cubic foot.	"	.28	"	.36
Blacksmith shop, brick, per cubic foot	"	.10	"	.13

Cost of heating, including hot water, boiler, pipes, radiators, valves, etc., complete for each 1,000 feet of cubic contents—

Churches	6.00
Hospitals, and similar buildings	16.00
Factories and mills	10.00
Dwellings, clubs, etc	21.00

These amounts include everything in connection with the heating except the boiler house.

Cost per cubic foot of houses built in good style of pressed brick facings, or fine stone, well finished in hardwood, oak, or birch	from	$0.30	to	$0.41
Brick buildings, of less pretensions, per cubic foot	"	.27	"	.38
Brick, third class, per cubic foot	"	.20	"	.30
Brick, fourth class, per cubic foot	"	.15	"	.25
Brick, fifth class, per cubic foot	"	.12	"	.21
Libraries, complete in brick, per cubic foot	"	.17	"	.25
Libraries, complete in stone, per cubic foot	"	.19	"	.30
Mortuary chapels, complete, per cubic foot	"	.25	"	.33
Museums and similar buildings, per cubic foot	"	.23	"	.34
Opera houses, first class, per cubic ft.	"	.30	"	.40
Opera houses, second class, per cu. ft.	"	.25	"	.35
Opera houses, third class, per cubic ft.	"	.22	"	.32
Opera houses, fourth class, per cu. ft.	"	.20	"	.28
Prisons, complete, including padded cells, per cubic foot	"	.18	"	.20
Cost per cubic foot for tearing down old brick buildings, including walls, chimneys, partitions, taking up floors, and removing window and door frames, sashes, doors and finishings, moving away debris, cleaning site and old materials and stacking up brick, joists, frames, lumber, etc. The whole cubic contents of building to be measured from bottom of footings to half-way up roof, per cubic foot	"	.01	"	.01½
Frame skating rinks, per cubic foot	"	.09	"	.12
Brick skating rinks, per cubic foot	"	.10	"	.13
Riding schools, with track, per cu. ft.	"	.13	"	.15
Sheds, rough, in wood, per cubic foot.	"	.05	"	.08
Sheds, rough, in brick, per cubic foot.	"	.08	"	.10

Sheds, rough, in iron, per cubic foot	from	$0.09	to	$0.12
Stores, dry goods, wood, per cubic foot	"	.13	"	.15
Stores, dry goods, brick, per cubic foot	"	.15	"	.17
Stores, dry goods, first-class finish, brick, per cubic foot	"	.20	"	.28
Stores, dry goods, second-class finish, brick, per cubic foot	"	.18	"	.24
Stores, dry goods, third-class finish, brick, per cubic foot	"	.16	"	.20
Stores, groceries, wood, good finish, per cubic foot	"	.14	"	.16
Stores, groceries, brick, fine finish, per cubic foot	"	.16	"	.18
Stores, groceries, brick, first-class finish, per cubic foot	"	.18	"	.22
Country or town halls, in brick or stone, well finished, classic style, with all necessary appointments and fittings, marble wainscot and other corresponding finish inside and out, per cubic foot	"	.32	"	.40
For country, per cubic foot	"	.30	"	.38
For cities, per cubic foot	"	.36	"	.42
For states, per cubic foot	"	.45	"	.55
For states, with towers, per cubic foot	"	.46	"	.57
Water towers, brick, per cubic foot	"	.16	"	.20
Water towers, iron, per cubic foot	"	.17	"	.20
Water towers, stone, per cubic foot	"	.20	"	.22
Model cottages, stone dressing, brick, per cubic foot	"	.13	"	.16
Model cottages, stone dressing, second class, per cubic foot	"	.12	"	.14
City flats, brick, per cubic foot	"	.28	"	.30
City flats, stone, per cubic foot	"	.30	"	.32
City flats, stone and brick, per cubic ft	"	.29	"	.31
Street arches for gala days, if of rough wood, covered with bunting, mottoes, evergreens, and similar materials, and are only temporary, per cubic foot	"	.04	"	.08

Better-class arches, plastered, etc., per cubic foot .... from $0.07 to $0.12
If made with staff and moulded, and have statuary, per cubic foot ... " .10 " .25
Permanent arches, in stone, per cu. ft. " .55 " 1.00
Permanent arches, first class, in marble, per cubic foot .... " 1.25 " 3.00

City parks—exclusive of land—walks, drives, lakes, buildings, roads, gates, walls, rustic bridges, and other things in connection with well-appointed parks, per acre—

First class	$3,000.00
Second class	2,500.00
Third class	1,800.00
Fourth class	1,000.00
Fifth class	600.00

Parks in country towns, or large villages where exhibition buildings, offices, and stables are kept, in conjunction with a race-course, and the area not less than twenty-five acres, the total cost of artificial work, including rough buildings, should not be more than, per acre $575.00

Cost of exhibition buildings, of wood,

First class, per cubic foot	from	$0.09	to	$0.11
Second class, per cubic foot	"	.06	"	.09
Third class, per cubic foot	"	.05	"	.07
Fourth class, per cubic foot	"	.04	"	.06
Fifth class, per cubic foot	"	.03	"	.05

Exhibition buildings for pigeons, cows, horses, sheep, poultry, etc.

First class, wood, per cu. ft.	from	$0.08	to	$0.10
Second class, wood, per cu.ft	"	.07	"	.09
Third class, wood, per cu. ft.	"	.06	"	.08
Fourth class, wood, per cu.ft.	"	.05	"	.07
Fifth class, wood, per cu. ft.	"	.03	"	.05

These items cover most of the ground for cubing, and are taken from the best authorities on the subject and from actual experience, and are quite sufficient for the ordinary purposes of the estimator who is likely to purchase this book.

As I have stated before, the cube rate cannot be relied upon for work of exceptional elaboration. The cubes generally published are intended to apply chiefly to buildings of a plain character in their several classes, and it would be of value if this circumstance were taken into account in fixing upon the rate. Precision can, however, only be attained by a generalization from extensive experience. The rates must be taken as general guides in forming an estimate of cost, and in all cases the experience of the expert estimator can alone give value to the system. There can be no comparison between a large block of stores and an elaborately fitted up hotel. The one is comparatively simple to the other; the decoration to the hotels in an avenue would alone increase the cost per cubic foot. The materials may be the same, brick or stone, with the same kinds of materials for finish, but the cost of labor, sizes of rooms, difference in walls, in heating, in plumbing, etc., would make a vast difference in the cost per foot, as an authority says on this subject: "I think the probabilities are that the cubing of a building 100 feet high would be higher than that of a building 50 feet high. It altogether must depend upon whether the larger building and the higher building has rooms of nearly the same size as the smaller building. No doubt the higher building would require thicker walls, but immediately you get away from comparatively small rooms into very large cubic spaces, then the difference in price is not great."

In fewer words we may say that the cost per foot cube of a building depends mainly upon the divisional internal walls and floors; the more numerous the rooms into which the space is divided, the greater the cost. Height is certainly a factor of cost, as a high building requires thicker walls; scaffolding and labor become expensive. But if we take two buildings, one twice the superficial area of the other, but of the same height, the difference per foot would entirely depend on the interior division and elaboration of plan. But to say that the cubing of a bigger and higher building is *pro rata* higher than for a smaller and lower one is a proposition that does not always hold. It is so only when the rooms are about the same dimensions in both cases. It would, for instance, be absurd to cube a large public hall with the usual rooms at a higher ratio than a small villa residence, because it was larger or higher. In plain English, the greater internal space and vacuities the less charge must be placed on the cube foot.

With regard to ornamental façades of wrought stone, a considerable addition per foot must be made upon the cost of a plain brick front. To cube both at the same figure would be wrong.

It may be asked, then, would any successful builder take a contract on the figures derived from cubing? We may answer that half the estimates now made by architects, in their private and public capacities, are made by cubing, and that contractors are to be found who would willingly take the risk of carrying out work in that manner. The two most perilous rocks upon which the cuber comes to grief are those of taking a figure without the verification of experience, and not making any allowance for internal elaboration of plan and decoration.

ESTIMATING BY DETAIL QUANTITIES

We now come to the only method on which the small contractor can depend, and which is always reliable if the estimator only does his duty properly and refrains from "jumping" at the prices, a trick many estimators employ to evade a little work in figuring.

I have given, in the first pages of this work, a detailed method of estimating for excavating, ditching, rough walling, concreting, and other like matters, to which the reader is referred when he is called upon to estimate on such work, so I will now make a departure and reproduce a system, corrected and brought up to date, which I published in *The Builder and Woodworker* of New York, in February, 1879, and which, in my opinion, has never been improved. The system was quite popular and many thousand copies of it have been sold. Insurance appraisers and others have made it a "text-book" to some extent, and used it with the adjustment of prices, of course, to suit the time and locality.

The list of items given in former pages must be followed, but there will be many others that will crop up which the estimator must provide for when preparing his tender, and these he should make a note of for future reference. It would be well to copy the items I have given in a good-sized book, leaving a generous margin for any remarks or notes it may be necessary to make, and new items should be entered as they appear.

We will suppose the building to be figured on is to be a balloon frame; the total cost of it can be closely calculated when the price of material and wages per day or hour are known.

First, mark on the plan, in plain figures, all the

dimensions and measurements in the building on which you are to estimate. Next, get the lineal measurement of all the sills, and from their size estimate the number of feet, board measure. Retain the lineal measurement, as from that the labor amount is estimated. The labor on the sills may be summed up to three kinds: First, framing without gains for joists or mortises, for studding as in common building when the studding is spiked to the sills and the joists rest on their top. Second, with mortises for studding, gains for joists, or studding without mortises. Third, with both mortises and gains.

Sills, 6 × 8, framed and placed in the building by the first, second and third processes, will cost for labor about 3, 5 and 7 cents per lineal foot. Sills, 12 × 16, double above prices. The intermediate sizes can be approximated from the above figures.

Joists are ordinarily placed 16 inches from center to center, and when so placed the number of joists on a given floor can be found by taking ¾ of the length of the building and adding one joist where they are placed on top of the sill, and deducting one where the end sills are used in place of joist. First floor joists usually are 2 × 8 to 2 × 14. Second floor 2 × 8 to 2 × 12. Ceiling joists, where no floor rests thereon, are 2 × 6 to 2 × 8.

Two men will frame and place in a wood building, not exceeding three stories, 600 lineal feet of joists, in size from 2 × 6 to 2 × 14 stuff, in one day of 8 hours.

In brick buildings not exceeding three stories, including anchoring and leveling up, 400 feet. Fourth story work, 350, and fifth story, 275 lineal feet.

The cost per lineal foot can be had from the above figures.

When joists are doubled under chimneys or partitions, the number of joists so used must be added to the result above named.

In balloon frames no braces are used. In timber frames they are made as follows:

1st. Cut off plain, spiked in, or "flat foot."

2d. With short tenons; and 3d, with long tenons and pinned. Braces vary in size from 4 x 4 to 6 x 6. The cost of labor will not vary on account of difference in size. The first pieces will cost 2 cents, the second 3½ cents, and the third 4½ cents per lineal foot, framed and placed in the building.

The plates in a balloon frame are made of scantling of the same size as the studding, and are worth to get out and spike to the frame 1½ cents per lineal foot.

In timber frames the labor on plates is: (1) framing without braces or gains for rafters; (2) framing with braces and no gains for rafters; (3) framing with both braces and gains. An average price for labor on plates in sizes from 4 x 6 to 6 x 10 would be: first process, 2½ cents; second process, 5 cents; third process, 7 cents per lineal foot. From 8 x 12 to 12 x 16, respectively, 4, 6 and 9 cents per lineal foot. This includes placing them in the building. Plates laid on walls are worth the same as plates spiked on the joists.

Posts in balloon frames are merely double-studding. The cost of placing them in position is the same as for studding.

Posts for timber frames are framed, first, with tenon top and bottom; second, the same, with one set of braces with girth or beam mortises; and third, the same, with two sets of girth or beam mortises.

By the first process posts from 4 x 6 to 8 x 10 would cost 4 cents. Second process, 6, and the third process, 9 cents per lineal foot to frame and place in the building.

Studding for balloon frames is usually placed 16 inches from center to center. They vary in size from 2 x 4 to 2 x 6. Occasionally odd sizes are used, as 2½ x 4, 2 x 5, or 3 x 4. In an ordinary size frame building two men will lay out and raise 800 lineal feet of 2 x 4 studding per day, or 750 feet of 2 x 6.

At $3 per day, the first would cost 77 cents per 100 lineal feet. The latter, 86 cents. The labor of spiking of joists and plates being considered under their respective heads, the work on studding is simply con fined to tenoning and studding on end, or spiking them to the sills.

A short rule for getting the number of pieces of outside studding, including plates, and allowing for doubling at all corners, and for windows and doors, is simply had by allowing one piece of studding for every foot of outside measurement.

This rule for buildings having many angles, where studding must be doubled, approximates very closely to the true result. In smaller buildings, without any angles, it will somewhat overrun.

The exact number of pieces of studding on the outside of building may be found by taking three-fourths of the number of feet in the outside measurement of the building; add one stud for each corner and angle, and one for each door and window. To this add for plate and gable studding.

Three-fourths of the number of lineal feet of all partitions will give the number of pieces required.

Their length, of course, depends upon the height of the rooms.

The cost of labor is the same as for outside studding.

It frequently happens that the studding is not double for doors and windows, and occasionally the extra stud for the corners in omitted.

Ribs for studding are usually made from 1 to 1½ inch stuff, and will cost to lay out and nail to the studding about 1 cent per lineal foot. The purpose for these is to support the upper joist.

Three-fourths of the width of the building, less one, gives the number of pieces required for gable; the average length of each piece is the distance from the plate to the ridge of the roof, or what is termed the rise of the rafter.

Rafters are designated as main or principal rafters, hip, jack, and valley rafters, and plain rafters.

The long rafters of a hip roof are called the main or principal rafters.

The shorter ones are called jack rafters.

A plain rafter is the ordinary rafter used in straight gable roofs.

The projection of a rafter is the distance it extends beyond the plate, or the length of the look-outs.

The *rise* of a rafter is the height on a perpendicular line from the plate to the ridge of the roof.

The *gain* of a rafter is the difference between the run and its length.

The run of a rafter is the distance from the outer edge of the plate to a point immediately under the ridge of the roof, or one-half the width of the building.

For a common rafter, to the square of the rise, add

the square of the *run*. The square root of their sum is the length of the rafter from the outer edge of the plate to the ridge of the roof.

The *rise* of a rafter is found by multiplying the number of inches rise required by the run by one-half the width of the building.

The *rise* in one-quarter pitch is one-quarter the width of the building. In a one-third pitch, one-third the width of the building. In a one-half pitch, one-half the width of the building, etc.

A common rafter can also be found as follows: If the roof is one-quarter pitch, to the square of one-quarter of the width of the building add the square of one-half the width of the building. The square root of the sum will be the length of rafter required. If a roof is one-third pitch square, one-third of the width of the building. If one-half pitch square, one-half the width, etc., and then proceed with the balance of the rule.

Required the length of rafters for a building 24 feet wide, gable roof, and one-quarter pitch.

One-fourth of 24 equals 6; ½ of 24 is 12. Squaring both gives 36 and 144, or 180; the square root of which is 13.416 feet, or length of rafter required.

Rule for estimating the length of rafters for hip roofs where they are of equal lengths:

Get the length of the main rafter by using the rule for common rafters. Then divide the length of the main rafter into one more space than the number of rafters required. The length of the space is the length of the shortest jack rafter, and the length of each studding rafter is simply the space added to the length of the preceding one.

Example.—Main rafter, 24 feet. Number of jack

rafters required, 7. Hence the number of *spaces* would be 7+1, or 8. Dividing 24 by 8 gives 3 feet as the length of the shortest rafter. The next would be 6 feet, then 9 feet, 12 feet, 15 feet, 18 feet, 21 feet, and then comes 24, or the main rafter.

Common rafters on shingle roof are placed from 16 to 24 inches from center to center, according to the length and weight of roof required; generally 2 feet is the distance.

The number of rafters in a plain gable roof is found by dividing the length of the building by the distance the rafters are apart from center to center, to which add 1; the result is the number of *pairs* of rafters.

Cost of Framing Rafters.—Two men in one day will frame and place in the building 600 lineal feet of 2 x 4 or 2 x 6 rafters—roof, plain gable.

In a hip roof, including framing for deck, if any, 250 feet is a fair day's work.

The former would cost 75 cents per 100 lineal feet, and the latter $1.75 per 100 lineal feet.

The contract price for framing one and a half, two, and two and a half story houses, in many of the Western states, averages 85 cents per 100 lineal feet of *all* the bill timber.

In all the framing labor thus considered, reference is had to soft wood only. If hard wood is used a fair addition to the prices would be 30 per cent.

If any of the work is circular, segment or octagonal, an addition must also be made, varying from two to four times the prices herein charged.

Lookouts for Hip Roofs.—An average length would be 20 inches. These are made of inch stuff and nailed to the rafters. They are worth, to get out, furnish material and place in position, 22 cents each.

The siding to a building is either drop siding, lap siding, dressed barn boards, or rough barn boards.

The number of feet of drop or lap siding is found by multiplying the outside measurement of the building by the height of the posts, to which add for gables, if roof is a gable roof, the product of the width of the building by the height from the plate to the ridge of the roof. This gives the number of surface feet, to which add one-fifth for lapping, and you have the number of feet board measure.

Two men will put on 700 feet in one day of drop siding when the window-casings and corner-boards are placed over the siding. Where joints are made against casings and corner-boards, 400 to 500 feet is a day's work.

Of lap siding, 650 feet. This includes putting up staging. Making the prices per square: Drop siding by the first method, 80 cents; second method, $1.20 to $1.50. Lap siding, 95 cents.

Two men will put on 2,000 feet of rough barn boards, or 1,500 feet of surfaced barn boards in one day, and will put on 2,000 feet of dressed battens, or 3,000 of rough battens. Hence the price would be: rough barn boards, 30 cents per 100 feet or one square; surface barn boards, 35 cents per 100 feet or one square. Dressed battens, 30 cents per 100 lineal feet. Rough battens, 18 cents per 100 lineal feet.

Roofs.—The area of a plain gable roof is had by multiplying the entire length of the rafters by the length of the building, including the projection of the cornice This gives one side; doubling it gives the total square feet of roof.

Hip Roofs.—Get the entire outside measurement of the building, including the projections of the cornice.

Multiply this by the length of the principal rafter and take one-half; the result is the area of the roof.

Hip Roof with Deck.—To the outside measurement of the deck, add the outside measurement of the building as above. Multiply this by the length of the principal rafter, and take one-half for the area of the roof.

Roof boards for plain gable roofs are worth 40 cents per square to put on the building, and for hip roofs 60 cents per square.

If roof boards are matched stuff for tin or slate roof, charge $1.00 per square for gable and $1.25 per square for hip roofs.

Shingles.—The average width of a shingle is 4 inches. Hence when shingles are laid 4 inches to the weather, each shingle averages 16 square inches; and 900 are required for a square of roofing.

If 4½ inches to one another, 800 will cover a square.
If 5 inches to one another, 720 will cover a square.
If 5½ inches to one another, 655 will cover a square.
If 6 inches to one another, 600 will cover a square.

This is for common gable roofs. In hip roofs, where the shingles are cut more or less to fit the roof, add 6 per cent to above figures.

A carpenter will carry up and lay on the roof from 1,500 to 2,000 shingles per day, or 2 to 2½ squares of plain gable roofing, so that an average price per square for simply laying the shingles would be $1.40. Add 40 cents for laying the roof boards, and the labor account on a common shingle roof would be $1.80 per square.

Tin Roofs.—A sheet of roofing tin is 14 × 20 inches, and a box of tin contains 112 sheets.

Allowing the usual amount for side ribs and top and bottom laps, a box of tin will cover 182 square feet, and is worth about $6.50 per box. 1 C. charcoal.

Laying a box of tin will cost as follows:

1 box 1 C. charcoal tin	$6.50
10 pounds solder, 15c	1.50
Preparing tin for roof	1.80
Laying tin, 1 1/5 days	3.20
Total	$13.00

Valleys.—Tin valleys for shingle roofs are generally 14 inches, and for slate roofs 20 inches wide. An average price put on the roof, including material, would be 12 cents per square foot. One man will lay 1½ squares per day of valleys, in plain work; when roof is steep or valleys cut up, 1 square is a day's work.

Flashings.—Tin flashings for chimneys and where one part of a building joins another are worth, put on, 13 cents per square foot.

Gutters and Spouts.—

Gutters, 4-inch, are worth, put up, 12 cents per lin. foot.
Gutters, 5-inch, are worth, put up, 14 cents per lin. foot.
Gutters, 6-inch, are worth, put up, 17 cents per lin. foot.
Down spouts, 2-inch, are worth, put up, 10 cents per lin. foot.
Down spouts, 3-inch, are worth, put up, 12 cents per lin. foot.
Down spouts, 4-inch, are worth, put up, 14 cents per lin. foot.
Down spouts, 6-inch, are worth, put up, 30 cents per lin. foot.

Slate Roofs.—The prices per square for slate roofs can be had of slaters in any of our towns and cities.

They will vary from $8 to $11 or $14 to $16 per square.

The following table will be found useful to the estimator.

SLATER: MEMORANDA

Names.	Size.	Gauge for 3 in. Lap nailed in center.	Gauge for 3 in. Lap nailed 1 in. from head.	No. of Squares covered by 1200.	Weight of 1200, First Quality.	No. required to cover one Square at 3 in. gauge.	Weight per Square, First Quality.	Nails required per Square. Iron.	Nails required per Square. Copper.
	in.	in.	in.		cwt.		cwt.	No.	lbs.
Singles	12 x 8	4½	4	3.0	18	400	6	800	5
Doubles	13 x 6	5	4½	2.5	15	480	6	960	6
Ladies.	16 x 8	6½	6	4.5	25	266	5½	532	3½
Viscountesses.	18 x 10	7½	7	6.2	35	192	6½	384	2¾
Countesses. . .	20 x 10	8½	8	7.0	40	170	5⅔	340	4
Marchionesses.	22 x 11	9½	9	8.7	50	138	5¾	276	3¼
Duchesses....	24 x 12	10½	10	10.4	60	115	5¾	230	3
Princesses....	24 x 14	10½	10	12.2	70	98	5⅔	196	3
Empresses ...	26 x 16	11½	11	15.2	95	79	6¼	158	3½
				A.					
Imperials.....	30 x 24	13½	—	2.5	—	36	8	72	3
Rags.	36 x 24	16½	—	2.2	—	25	9	50	3½
Queens.......	36 x 24	16½	—	2.2	—	25	9	50	3½

A.—Squares covered by 1 ton.

The above sizes sometimes slightly vary, according to the quarry.

Slates are classed according to their straightness, smoothness of surface, fair even thickness, presence or absence of discoloration, etc. They are generally divided into first and second qualities, and in some cases a medium quality is quoted. Slates of first quality are thinner and lighter than those of inferior quality.

Rule to find the number of slates required to cover one square: One square in inches ÷ width of slate in inches × gauge in inches.

The weight of slating on roofs is 8 pounds per foot super. for all sizes, except rags or queens, including a 3-inch lap and nails.

As there are two nails per slate, the number required per square will be found by doubling the number of slates. The trade "thousand," or "long tally," equals 1,200 for buying and selling.

Nails.—Composition nails are best for all good work, as they are stiff and tough. They are cast from an alloy of 7 copper to 4 zinc, and have a yellow, brassy appearance. Copper nails are either cast or wrought; but they are soft and dear. Malleable iron nails are frequently used, dipped while hot in boiled linseed oil to preserve them from corrosion. These can also be painted or galvanized. Cast-iron nails are only employed for temporary work. Zinc nails are very soft, and liable to bend, and as their heads come off in driving, they make a good deal of waste.

All these nails are sold by weight, and the price should lessen with the increase of length. Allow 5 per cent for waste in reckoning the number to the square.

Nails for small slates, such as Doubles, etc., should be about	1¼ in. long
Nails for medium slates, such as Countesses, etc., should be about	1½ in. long
Nails for large slates, such as Duchesses, etc., should be about	2 in. long

SLATE NAILS

Galvanized slate nails, per keg, 3d	$5.50
Galvanized slate nails, per keg, 4d	5.00
Tinned slate nails, per keg, 3d	5.75
Tinned slate nails, per keg, 4d	5.25
Polished steel wire nails, 3d and 4d	4.00
Copper slate nails, per pound	.20

These prices vary with time and locality.

Labor.—The labor in holing slates, any size, is usually estimated at $1.50 per thousand; but if a single slate-holing machine is used, a smart boy, at 15 cents per hour, will be able to hole from 300 to 400 slates in an hour.

The following statement shows the labor required per square, which will be less for larger surfaces, as the slating will be performed more quickly. The difference in time for the various kinds represents the extra trouble in handling, greater areas being covered with larger slates in a given time, and the labor in holing is the same for all sizes.

A slater and assistant will lay:—

1 square of	Doubles (with two nails each)	in 2½	hours.
"	Ladies " "	" 1½	"
"	Countesses " "	" $1\frac{1}{10}$	"
"	Duchesses " "	" 1	"

A slater and assistant will prepare and lay:—

1 square of	Doubles (with two nails each)	" 4	"
"	Ladies " "	" 2½	"
"	Countesses " "	" 2	"
"	Duchesses " "	" 1½	"
Plastering against underside of slating, per yard super.		" ½	"

Cost per Square.—Taking Countess slates, 20 inches long by 10 inches wide, the gauge, if center-nailed, would be: $\frac{\text{Length of slate} - \text{lap}}{2} = \frac{20 \text{ in.} - 3 \text{ in.}}{2} =$ 8½ inches. In estimating, therefore, the number of slates required per square of 100 feet super., the width of the gauge in inches, multiplied by the breadth of the slate in inches, gives the margin or exposed surface of a single slate. This divided into the number of superficial inches in a square (100 feet super. by 144 square inches = 14,400 super. inches per square), will give the number of slates to a square—

e.g., 8½ inches gauge by 10 inches breadth of slate = 85 square inches margin, and $\frac{14,400 \text{ super. in. per square}}{85 \text{ sq. in. margin per slate}}$ = 170 Countess slates per square.

Allowing 5 per cent for waste, this would give roundly 180 slates to the square.

As there are two nails per slate, the number of nails required per square will be found by doubling the number of slates—i. e., in this case, 340 nails. Also reckoning 5 per cent waste for nails, the number for estimating would be some 360. Using 1½-inch composition nails, 144 of which go to the pound, this latter number would give exactly 2½ pounds per square, as they are sold by weight.

A slate roof is laid by first placing a course on the eaves. All courses above this one must be laid with a lap of more than one half the length of the slate or the vertical joints which are not close will not be covered. The lap of the slate is more than one-half its length, so the more lap a course is laid with, the better will be the roof. Manufacturers allow 3 inches when selling a square of slate, and architects and consumers should see that the roof is laid with that amount of lap, as a less one is a considerable gain for the dishonest roofer, which he takes advantage of to the permanent injury of the roof, because any less lap than 3 inches greatly endangers the weather-proof qualities of a slate roof. Slate, before it is laid, should be carefully sorted, the thick ones used to start the roof at the eaves and the thin ones to finish with at the comb. In nailing slate do not drive the nails too tight. The top of the nail should be just even with the surface of the slate.

NUMBER OF SLATE IN ANY NUMBER OF SQUARES, FROM ½ UP TO 60 SQUARES

Size	½ Sq	1 Sq.	2 Sq.	3 Sq.	4 Sq.	5 Sq.	6 Sq.	7 Sq.	8 Sq.	9 Sq.	10 Sq.	11 Sq.	12 Sq.	13 Sq.	14 Sq.	15 Sq.	16 Sq.	17 Sq.	18 Sq.	19 Sq.	20 Sq.	30 Sq.	40 Sq.	50 Sq.	60 Sq.
24 x 16	43	85	171	258	343	428	515	600	685	772	857	943	1029	1115	1200	1286	1372	1458	1543	1629	1715	2572	3449	4286	5143
24 x 14	49	98	196	294	392	490	588	686	783	881	979	1077	1175	1273	3171	1469	1567	1665	1763	1861	1959	2938	3918	4897	5877
24 x 12	58	115	229	343	457	571	686	800	914	1029	1143	1257	1371	1485	1600	1714	1828	1942	2057	2171	2285	3428	4571	5714	6857
22 x 14	54	108	217	325	434	542	650	758	866	975	1083	1191	1300	1408	1516	1625	1733	1841	1949	2058	2166	3294	4331	5414	6497
22 x 12	63	126	253	379	505	631	758	884	1011	1137	1263	1389	1515	1642	1768	1894	2021	2147	2273	2400	2526	3789	5052	6315	7578
22 x 11	69	137	276	413	551	689	826	965	1102	1240	1378	1515	1653	1791	1929	2066	2204	2342	2480	2618	2755	4133	5511	6889	8267
20 x 14	61	121	242	363	484	605	726	847	968	1089	1210	1331	1452	1573	1094	1815	1936	2057	2178	2299	2420	3630	4840	6050	7260
20 x 12	71	141	282	424	565	706	847	988	1129	1271	1412	1552	1694	1835	1976	2117	2258	2400	2541	2682	2823	4235	5647	7058	8471
20 x 11	77	154	308	462	616	770	924	1078	1232	1386	1540	1694	1848	2002	2156	2310	2464	2618	2772	2926	3080	4620	6160	7700	9240
20 x 10	85	170	339	508	678	847	1017	1186	1356	1525	1694	1863	2032	2202	2371	2541	2710	2880	3049	3218	3388	5082	6776	8470	10164
18 x 12	80	160	320	480	640	800	960	1120	1280	1440	1600	1760	1920	2080	2240	2400	2560	2720	2880	3040	3200	4800	6400	8000	9600
18 x 10	96	192	384	576	768	960	1152	1344	1536	1728	1920	2112	2304	2496	2688	2880	3072	3264	3456	3648	3840	5760	7680	9600	11520
18 x 9	107	213	426	640	853	1066	1280	1493	1706	1920	2133	2346	2560	2773	2986	3200	3413	3626	3840	4053	4266	6400	8533	10666	12800
16 x 12	93	185	370	554	739	924	1108	1293	1477	1662	1847	2031	2216	2400	2585	2770	2954	3139	3324	3508	3693	5539	7385	9230	11077
16 x 10	111	222	443	664	886	1107	1329	1550	1772	1993	2215	2436	2658	2880	3101	3323	3544	3766	3987	4209	4430	6646	8861	11076	13292
16 x 9	123	246	492	738	985	1231	1477	1723	1969	2215	2461	2707	2953	3200	3446	3692	3938	4184	4430	4676	4928	7384	9846	12307	14769
16 x 8	138	276	554	831	1108	1385	1662	1938	2215	2492	2769	3046	3323	3600	3876	4153	4430	4707	4984	5261	5538	8307	11076	13846	16615
14 x 14	94	187	374	561	748	935	1122	1309	1496	1683	1870	2057	2244	2431	2618	2805	2992	3179	3366	3553	3740	5610	7480	9350	11220
14 x 12	109	218	437	654	872	1091	1310	1527	1745	1963	2182	2400	2618	2836	3054	3272	3490	3709	3927	4185	4364	6545	8727	10909	13090
14 x 10	131	262	524	785	1048	1309	1570	1833	2094	2356	2618	2880	3141	3403	3665	3927	4189	4450	4712	4974	5236	7854	10472	13090	15709
14 x 9	145	290	581	872	1163	1454	1745	2036	2326	2618	2909	3200	3490	3781	4072	4363	4654	4945	5236	5527	5820	8727	11636	14545	17454
14 x 8	164	327	655	982	1309	1636	1964	2291	2618	2940	3273	3600	3927	4254	4581	4909	5236	5563	5890	6217	6545	9818	13090	16363	19636
14 x 7	187	374	748	1122	1496	1870	2244	2618	2992	3366	3740	4114	4488	4862	5236	5610	5984	6358	6732	7106	7480	11220	14961	18701	22441
12 x 12	134	267	534	800	1067	1334	1600	1867	2133	2400	2667	2934	3200	3467	3734	4000	4267	4534	4800	5067	5334	8000	10667	13334	16000
12 x 10	160	320	640	960	1280	1600	1920	2240	2599	2879	3200	3520	3840	4160	4480	4800	5120	5440	5760	6080	6400	9600	12800	16000	19200
12 x 8	200	400	800	1200	1600	2000	2400	2800	3200	3600	4000	4400	4800	5200	5600	6000	6400	6800	7200	7600	8000	12000	16000	20000	24000
12 x 7	229	457	914	1371	1828	2285	2743	3200	3657	4114	4571	5028	5485	5942	6399	6857	7314	7771	8228	8685	9142	13714	18285	22857	27428
12 x 6	267	533	1067	1600	2134	2667	3200	3734	4267	4800	5334	5867	6400	6934	7437	8000	8534	9167	9600	10134	10667	16000	21334	26667	32000

Cornices.—An ordinary plain cornice has three members, viz.: frieze, soffit, and fascia.

The frieze is the part nailed or fastened to the side of the building.

The soffit is the part attached to the under side of the projection of rafter, or lookout.

The fascia is the part attached to the end of the rafters or lookout.

Crown moulding is the moulding on the fascia.

Bed moulding is the moulding in the angle where the frieze and soffit join.

In estimating the amount of material in a given cornice for a square roof, multiply the entire outside measurement of the building by the sum of the width of the soffit, frieze and fascia; the result is the number of feet, board measure.

For gable roofs, to the lengths of the two sides of the building add the end projections and length of end rafters and multiply as before.

Table of labor account on cornice work.

Number of feet two men will put on per day and price per foot:

Width in Inches				
Frieze	Soffit	Fascia	No. Feet	Cost per foot
9	10	4	80	7½
10	12	4	75	8
12	16	4	60	10
14	20	5	48	12½

The above is for gable roofs and includes cost of scaffolding.

Hip Roofs.—

Frieze	Soffit	Fascia	No. Feet	Cost per Foot
18-inch.	16-inch.	4-inch.	75	8
22 "	20 "	4½ "	64	9½
28 "	24 "	5 "	52	12
32 "	28 "	5½ "	40	15
34 "	32 "	6 "	32	20

Cornice Mouldings.—

Crown moulding, flat,		2-inch.	800 feet per day,		or 80c.	per 100 feet.
"	"	spring 4	"	500	"	$1.20	"
"	"	" 5	"	445	"	1.31	"
"	"	" 6	"	365	"	1.62	"
"	"	" 7	"	300	"	2.00	"
"	"	" 8	"	250	"	2.40	"

The cost of cornice moulding is ordinarily 1 cent per lineal foot less than the number of inches in work —2-inch moulding, 2 cents; 3-inch, 3 cents, etc.

Bed moulding, flat, 1½-inch, 800 feet per day, or 80 cents per 100 feet. Bed moulding, flat, 2-inch, 750 feet per day, or 84 cents per 100 feet. Bed moulding, flat, 3-inch, 700 feet per day, or 88 cents per 100 feet. Bed moulding, flat, 4-inch, 500 feet per day, or $1.20 per 100 feet.

Cornice Brackets.—Price per bracket, soft wood, all well worked—cost to put on building:

Perpendicular	Horizontal	Thickness	Cost Plain	Moulded	Plain	Moulded
Size, 16-inch.	12-inch.	2½-inch.	$0.35	$0.42	$0.15	$0.20
" 20 "	16 "	3 "	.70	.80	.20	.25
" 24 "	20 "	4 "	.70	.85	.14	.20
" 28 "	24 "	5 "	1.00	1.20	.25	.35
" 30 "	28 "	6 "	1.50	1.60	.35	.45

Plain panel moulding, two men will put on 300 feet per day. Foot moulding, two men will put on 400 feet per day.

FLOORS

						Cost per Square
Soft wood,	6 in. wide,	without bridging,	per joist,	800 sq. ft.		$0.80
"	6 "	with "	"	650	"	.90
"	4 "	without "	"	600	"	.98
"	4 "	with "	"	500	"	1.04
"	3½ "	without "	"	400	"	1.25
"	3½ "	with "	"	300	"	1.50

Two men will dress six squares of flooring after laying per day, or at a cost of $1.00 per square.

If flooring is of hard wood, estimate per day two-thirds of above.

The number of feet, board measure, in a given floor is had by multiplying its length by its width and adding one-fifth for lapping. For flooring not matched omit the lapping. Two men will lay 1,333 feet of plank flooring per day, or 45 cents per square, or will lay 2,000 feet of common rough flooring, 1-inch stuff, or 30 cents per square.

Outside ceiling for wood buildings, average width, including beading and scaffolding, is worth, to put up, $1.25 per square. An average day's work for two men is five squares. Two men will dress, after laying the ceiling, five squares per day, or $1.20 cents per square. Ceiling overhead is generally of wider stuff than outside ceiling; as there is no beading, and the workmanship is not so particular, two men will put up the same amount as of outside ceiling, including putting up and taking down scaffolding, or five squares at 80 cents per square.

Wainscoting.—Wainscoting 2½ to 3 feet high, beaded, with ordinary capping, including dressing after putting up, is worth $3.00 per square. Two squares is a day's work for two men.

The same, 3 feet to 4 feet high, is worth, to put up, $2.00 per square.

The same, with shoe and heavy caps, is worth $2.60 per square. The capping to wainscoting is ordinary moulding from 1½ inches by ⅞ to 2 inches by 1⅛ inches.

Panel wainscoting, mill worked, ready to put up, including capping, shoe or base, is worth, for labor, $3.25 per square.

Hand-worked panel wainscoting is of so various a kind that definite prices of labor cannot well be given without specifications. In a general way, the price per square for getting out and putting up will vary from $3.00 to $20.00 per square.

The above prices are for soft wood. For hard wood add 40 per cent.

Baseboards.—Plain base, 6 to 10 inches wide, put up before plastering, is worth 1½ cents per lineal foot for labor. Two hundred feet is a good day's work for a man with mill-dressed lumber.

The same, put on after plastering, including putting on grounds, is worth 2 cents per lineal foot.

Plain base, after plastering, with moulding, leveling, or capping by hand—mill-dressed stuff—is worth 2½ cents per lineal foot to get out and place in the building.

Stairs.—The wall string is the board with which the ends of the steps are fixed next to the wall.

The face string is the board that carries the oute· end of the steps and risers.

The *tread* is the horizontal board of the step.

The *riser* is the upright board of the step.

The *newel post* is the upright post at the lower step to receive the hand rail.

The hand rail is the rail supported by balusters. Balusters are small columns or pillars to support the rail.

The number of risers is found by dividing the distance from floor to floor by the height of the rise.

The height of each rise is found by dividing the distance from floor to floor by the number of risers.

The number of treads is one less than the number of risers.

The width of each tread is found by dividing the risers by the number of treads and adding the projection.

Risers vary in height from 4 to 8 inches. Treads run from 8 to 14 inches:

It will be impracticable to give detail prices for all variety of stair-work on account of the diversity of designs. We simply give a few as an illustration. The labor on rough, open stairs, for cellars or stables, when no risers are used, is worth 16 cents per tread. Straight stairs between partitions, 2 feet 6 inches to 3 feet 6 inches long, with 6-inch to 9-inch tread, and 7-inch to 8-inch risers, are worth 35 cents per riser.

Winding stairs, same dimensions, 40 cents per riser. Open straight stairs, risers 6½ to 8 inches, treads 6 to 11 inches; housed in wall strings, mitered to face string; moulded nosing, including putting up turned balusters, and plain round or oval rail, with 6-inch to 8-inch turned newel post, are worth for labor $1.10 to $2.00 per riser.

The same stairs, winding, charge $2.50 per riser for the winding steps, and $1.25 for straight steps. Putting on brackets outside of stringer is worth from 5 to 12 cents per bracket.

The following is a list of the approximate prices of stair material:

Newel Posts.—A turned newel post of cherry or black walnut, 5 inches in diameter, with cap, is worth $3.50; 6 inches, $4.00; and 8 inches, $5.50.

Octagon newel posts, walnut, oak, or cherry, with ornamental cap, 8 inches, $8.00; 9 inches, $8.50; and 12 inches, $10.50.

Newel posts veneered with fancy woods, with carving on plinth and cap, and moulded sunk panels, will vary from $20.00 to $60.00 each.

Balusters.—Turned balusters, walnut or cherry, from 2 feet 4 inches to 3 feet, are worth, 1½ inches, 10 cents;

2 inches, 14 cents; and 2½ inches, 20 cents each. Oak and ash 20 per cent less.

Fluted or octagon balusters, walnut or cherry; 2 inches, 18 cents; 2½ inches, 25 cents; 2¾ inches, 30 cents each. Fancy balusters for high-priced stairs may run from 40 to 60 cents each.

Rails.—Walnut or cherry, 3½-inch, 15 cents; 4-inch, 20 cents; 4½-inch, 22 cents; and 5-inch, 22 cents per lineal foot. Raised back rails, walnut or cherry, 4-inch, 25 cents; 5-inch, 30 cents; 5½-inch, 36 cents; and 6-inch, 40 cents per lineal foot. Fancy raised back rails from 6 to 7 inches will vary from 50 to 70 cents per foot.

Doors.—The price of doors may be had from any dealer's catalogue. The labor account is as follows: A fair day's work for one man is setting 5 door frames a day, and putting on ordinary casing. He will also hang and finish 5 doors per day, or $1.20 a door complete. The above is for 6 feet to 7 feet 6 inch doors, and 1⅜ inch thick. From 7 feet 6 inch to 9 feet doors and 1¾ inch thick, a day's work of setting and casing 3 frames per day, or hanging and finishing 3 doors per day, $2.00 per door complete.

Moulding Door Casings.—For 6 feet to 7 feet 6 inch doors, and 3-inch mouldings, one man will mould 6 door casings, two sides, per day, or 50 cents per door; with 4½-inch mouldings, 5 doors per day, or 60 cents per door. Mouldings with two members about one-half above number, 7 feet 6 inches to 9 feet doors, single moulding two sides, 5 openings per day. The same, with double members to moulding, 2½ openings per day.

Door frames when had from factory are cased both sides for inside doors, and one side for outside doors.

Sliding Doors.—The frames for a pair of sliding doors with double joint, including casings each side, are worth from $3.50 to $4.00 per frame.

The same, with segment top, will vary from $6.00 to $9.50; setting either one of the above frames, putting up the track, and lining the pocket is worth from $3.50 to $4.00 for labor. Setting, hanging, and trimming a pair of sliding doors will take a man about 1¼ days, or $3.75 per door.

Folding Doors.—The frame for a pair of folding doors with opening 5 feet by 8 feet 6 inches, with single joints, including casing each side, is worth from $3.50 to $4.25 per opening. Segment top, same size opening, $6.00 to $8.00. Setting the frame for a pair of folding doors will take a man three-quarters of a day, or $2.25 per frame.

Fitting, hanging, and trimming a pair of folding doors will take one man a day and a quarter, or $3.75 per door.

Moulding, sliding and folding door casings, square top opening 5 feet by 8 feet 6 inches on both sides, single member; a day's work is 4 openings per day, or 75 cents per door. If moulding is double member, two openings per day, or $1.50 per door. Segment top with same size of swing, the moulding will cost $3.00 per opening. Over the face of a square top, one man will put on the moulding with a single member in one-half a day, or $1.50 per opening. Double member one day, or $3.00 per opening.

Setting door frames in brick buildings will cost the same as for frame buildings.

Common Door Frames.—Outside frames, with casings on one side for doors, from 2 feet 6 inches by 6 feet 6 inches to 2 feet 8 inches by 6 feet 8 inches, are worth from $2.25 to $3.50 each. The same for inside doors, with casing on both sides, are worth from $3.00 to $4.00.

Door Trimmings.—Butts 3 × 3 inches, for cheap trimmings, are worth 10 cents per pair, and a common mortise or rim lock, with brown knob, 30 cents each; 3 × 3½ butts, 10 cents, and 3½ × 3½, 10 cents each; 4 × 4, 15 cents. A good mortise lock, with brown or white knobs, brass key, face, and bolt is worth 45 cents. Outside door locks vary from 50 cents to $2.00 a pair; average price would be $1.00.

Sliding door locks 4 × 5, brass key and face, $1.50 each. Iron track for door, 3 cents per foot; brass track, 25 cents. A very good rabbeted lock, without night works, $1.50; with night works, $2.50 to $4.00 each.

Screws for putting on above trimmings, 30 cents a gross. The labor account for trimming doors will be found under the head of doors.

Windows.—The price of the sash, including glass and glazing for all sizes of windows, may be had from the dealers' catalogues. Window frames, factory made, simply have outside casings and jambs. One man will cut the openings and set five frames per day, of an average size, say 2 feet 6 inches by 6 feet, in a frame building, and can set the same number in a brick building, or 60 cents per opening.

As the brick-work goes up the carpenter must plumb up the frames occasionally, so that a fair estimate would be both alike.

In larger openings, setting from two to four frames per day would be fair work, or from 60 cents to $1.25 per window.

One man will case 12 windows per day of windows 2

feet 6 inches by 6 feet, or 1½ cents per lineal foot of the casing.

Moulding window casings, same price per foot as door casing.

For wood buildings, plain rail sash, 8 or 12 lights, with outside casings, an average price would be as follows:

8 × 10, $1.20; 10 × 12, $1.50; 10 × 14, $1.80; 10 × 16, $2.20.

With check-rail sash outside, casings: 8 × 10, $1.80; 10 × 12, $2.00; 10 × 14, $2.20; 10 × 16, $2.40; 10 × 18, $2.60.

Plain window frames for brick buildings: 8 × 10, $2.00; 10 × 12, $2.20; 10 × 14, $2.50; 10 × 16, $2.60; 12 × 24, $3.65.

Box window frames: 8 × 10, $2.85; 10 × 12, $3.00; 10 × 14, $3.20; 10 × 16, $3.50; 12 × 24, $4.30.

The same frames, with segment outside and square inside, are worth 50 cents more.

Pantries and Closets.—In ordinary work of this kind one man will get out and put up 50 to 75 lineal feet of shelving 12 inches wide per day, or will make and put up five drawers 15 inches wide by 18 inches deep, including racks and fitting.

If the drawers are dovetailed, four is a day's work. Strips and hooks: one man can put 50 to 80 lineal feet of strips, and put on closet hooks, about 12 inches apart, in one day.

Porches.—These differ so widely in design that prices per foot lineal cannot be given without specifications, as they will vary from $1.25 a foot upwards. In an ordinary porch, figure the sills and joists as in framing; also roof, labor, ceiling, and cornice the same as in other parts of the building, and charge for whatever extra work the design may call for.

Blinds.—These are made and sold by the foot, measuring height of the window on one side only; 60 to 70 cents per lineal foot, including trimming and hanging, is a fair price. Inside blinds, O. G. panel or rolling slats, ordinary width, are worth $1.25 per foot, complete in the building. If inside blinds are of hard wood, they are worth from one and a half to double the price of pine.

Plastering.—The number of yards is simply the area of all the walls and ceilings.

One hundred yards of plastering will require 1,400 laths, 4½ bushels of lime, 18 bushels of sand, 9 pounds of hair, and 5 pounds of nails for two-coat work.

Three men and one helper will put on 450 yards, in a day's work, of two-coat work, and will put on a hard finish for 300 yards.

Retail cost of three-coat work for 100 yards of plastering:

Seven bushels of lime at 30 cents	**$2.10**
Four-fifths of a load of sand at $1.25	**1.00**
Nine pounds of hair at 2⅔ cents	**.24**
Five pounds of nails at 4½ cents	**.22**
Lathing, 100 yards at 2¼ cents	**2.25**
1400 laths at $3.00 per 1000	**4.20**
Plastering, 2 coats, 1 man ⅔ of a day	**2.00**
Helper, ⅓ of a day	**.33**
Hard finished, one day's work	**3.00**
Making mortar and scaffolding	**1.50**
Total cost	**$16.84**

Or, say seventeen cents per yard.

Painting.—Painting is done by the yard, and at the present prices of lead and oil, house painting in plain colors will cost on an average:

For one coat, 8 cents per yard; two coats, 15 cents per yard; three coats, 23 cents per yard.

One coat, or priming, will take for 100 yards of painting 20 pounds of lead and 4 gallons of oil. Two-coat work, 40 pounds of lead and 4 gallons of oil. Three-coat, the same proportion; so that a fair estimate for 100 yards of three-coat work would be 60 pounds of lead and 12 gallons of oil.

A day's work on outside of a building is 100 yards of first coat, and 80 yards of either second or third coat. An ordinary door, including casings, will on both sides make 8 yards to 10 yards of painting, or say, 5 yards to a door without the casings. An ordinary window 2½ to 3 yards. Fifty yards of common graining is a day's work for a grainer and one man to rub in.

In measuring up outside work, use the rule for plain surfaces. In common painting run your tape-line over all the mouldings in and out, and this, with the width of the cornice multiplied by its length, will give the area. It is customary to add from one-third to one-half for the bracket painting. In painting blinds of ordinary size, twelve is a fair day's work for one coat, and 9 pounds of lead and 1 gallon of oil will paint them. In measuring up inside base, it is customary to reckon 9 inches in width and upwards to 1 foot as 12 inches.

Nails.—One thousand feet of inch stuff will require 10 pounds of 10-penny nails; 1 square of siding or ceiling, 2½ pounds 8-penny, and the same for a square of roof boards or sheathing, and 1,000 shingles will take 6 pounds of shingle nails.

Brick and Stone Work.—A day's work in excavating and filling into cart or wheelbarrow is 11 or 12 cubic yards of common earth, or 7 to 8 yards of clay or coarse gravel, or 14 to 16 cents per yard. In limestone or sandstone a day's work in quarrying will range from one-half to one cord of stone.

Stone Work.—A perch is 16½ feet long, 1½ feet wide, and 1 foot high, and contains 24¾ cubic feet. In estimates 25 cubic feet is figured as a perch.

A perch in the wall contains about 22 cubic feet of stone and 3 cubic feet of mortar.

The waste ordinarily allowed in laying stone walls from the rock measurement is one-fifth.

A cubic yard of rubble masonry laid in the wall contains 1⅛ cubic yards of undressed stone and one-fourth of a cubic yard of mortar.

Four perches or 100 cubic feet of wall will contain ordinarily 1 cord of stone or 128 cubic feet, 1 barrel of lime, or say 2½ bushels, and 5 barrels of sand.

A day's work for a mason's helper is moving 4 to 5 perches of stone, and mix and carry to the mason sufficient mortar to lay them.

A man will lay in one day from 4 to 5 perches of rubble masonry in sandstone, or 3 perches in limestone. In many locations sandstone is delivered for $1.25 per perch, and the labor for laying in ordinary walls, including lime and sand, from 95 cents to $1.25 per perch.

Stone Ashlars.—These are ordinarily 3 feet to 5 feet long, 1 foot high, and 4 to 6 inches thick.

The price of the rough stone will vary according to locality. The labor on ashlars, including setting, is per square foot as follows:

Fine posts, hammerwork,	limestone,	30 cts.;	sandstone,	25 cts.
Medium	"	"	28 "	"	22 "
Rough	"	"	20 "	"	17 "

Freestone ashlars, sawed, are furnished at the mills for 25 to 35 cents per square foot, and caps and sills for ordinary windows and doors from $1.35 to $1.70 each.

Brick Work.—The labor and material of brick work

are estimated by the 1,000 brick. In measuring up brick walls it is not customary to deduct for openings. To ascertain the number of bricks in a wall: First obtain the number of superficial feet, and multiply this by 7 for a 4-inch wall, by 14 for a 9-inch wall, 22 for a 14-inch wall, and 29 for an 18-inch wall. If thicker than 18 inches, for each additional 4½ inches in thickness add 7 bricks per square foot.

One thousand five hundred brick is an average day's work for outside and inside walls, and we take three-quarters of a barrel of lime and 9 bushels of sand to make the mortar. The number of brick a mason will lay in a day on a plain wall depends largely upon its thickness. On 9-inch work, 1,200 to 1,400; on 14-inch work, 1,500 to 2,000, and on 18-inch work, 2,000 to 2,500; veneered work or single-back walls attached to wood work is much slower, from 400 to 600 brick is regarded a day's work; this includes tying the brick with nails to the framework, or sheathing.

The following is given as an illustration of the cost of furnishing and laying 1,500 brick, or one day's work.

1500 brick at $6 per M	$9.00
¾ barrel of lime at $1	.75
9 bushels of sand at 5 cents	.45
1 day's work for mason	3.00
1 day's work for helper	2.00
Total	$15.20

Or, $10.14 per M.

Chimneys.—Common flues and ordinary chimneys are worth from 40 to 75 cents per running foot, including labor and material. In large chimneys with fire-places, get the number of brick, charge for lime and sand the same as in brick walls, and estimate the labor at double the price of plain walls of same thickness.

Plumbing.—In plumbing for bath-rooms and closets 1¼-inch pipe is used for water, ⅝-inch for supply, and 4-inch iron pipes for soil-pipe. An average price would be for material and putting in the building: 1¼-inch pipe, lead, 40 cents per foot; ⅝-inch pipe, lead, 32 cents per foot, and soil-pipe, 35 cents per foot.

Bath-tubs will vary in price from $15.00 to $50.00; double bath-cocks, $12.00 to $15.00; single, $1.90 to $3.00; wash-bowl cocks, from $2.00 to $3.00.

A fair price for a corner wash-bowl, marble, with stop-cocks and enclosed with casings, including connections with pipes, will vary from $12.00 to $20.00; water-closet basins and connections, $6.00 to $8.00.

It must be understood that the foregoing prices are only approximately correct.

SOME PAINTER'S EXTRAS

In estimating the painter's work, a few facts and data as to the quantity of paint required to cover certain areas of surface are necessary. Thus it is useful to know that 1 pound of mixed white lead paint will cover about 4½ superficial yards the first coat, and about 6½ yards each additional coat; that 1 pound of mixed red lead paint will cover about 5½ yards super. of iron. Some authorities say 45 yards of first coat, including stopping, will require 5 pounds of white lead, 5 pounds of putty and 1 quart of oil; and 45 yards of each succeeding coat will require 5 pounds of white lead and 1 quart of oil. These quantities do not exactly agree, but they are approximately correct, and we may take about 6½ to 7 yards to be about a fair allowance for 1 pound of paint; if the paint costs, say, 15 cents per pound, the cost would be about 2½ cents per yard for material; 1 pound of mixed white lead paint will cover 1 yard super. on Portland cement (first coat); good oil varnish requires 1 pint to 8 or 9 yards superficial, one coat.

In measuring the painting of iron railing, the two sides are measured as flat work, both sides plain, and charged as such, unless gilded; if the railing is delicate and ornamental, the charge is once and a half, or twice is taken for each side.

The rotation in taking the items are generally the windows, base dades, chimney pieces, doors; but this rule is not strictly observed, and in the abstracting the one-coat work comes before the three, four, or five times in oil; flatting and ornamental work follow the plain painting.

It may be useful to remember that the decimal .27 multiplied by the rate of wages for a painter per hour will give the cost per yard for common work, including stopping, knotting, etc., and the decimal .15 for second and following coats.

Staining, sizing and varnishing taken at per yard superficial should be described as to stain and the number of coats of varnish. For varnished work, state if on natural wood or painted. Graining and varnishing at per yard is similarly measured to plain painting, and should be described as "extra"; state if "combed," "once grained," and varnished, and the wood to be imitated as oak, walnut, etc., if once or twice varnished, and if with spirit or copal, if the wood is to be sized.

WOOD AND IRON WORK

95 yards 5 feet super. Knotting, stopping, priming, and painting wood work three times in oil and lead color. Taking the decimal .27 and multiplying by

rate of wages per hour would give, the cost per yard.

The price-books give 20 to 25 cents per yard for three-coat work.

103 yards super. Ditto four times on cement work. Add to the above 5 cents per yard, say, 22 cents per yard for a large quantity.

54 yards super. Painting four times balusters of staircase. These are ornamental and close, and the quantity given includes double face. Say, 25 cents per yard.

75 yards 6 feet super. Ditto five times iron railing. About 5 cents per yard more than last.

75-foot run. 4½ inch reveals in five oils. Worth about 8 to 12 cents per foot.

36-foot run. Painting r. w. pipes in four oils. Put this at 10 cents per foot.

66-foot run. Ditto eaves gutters. Same price.

35-yard run. Painting bars to skylights, four coats in oil. This is worth about 9 cents per yard.

120-foot run. Shelf edge, three coats. 3 cents per foot.

18-foot run. Painting in three oils, cornice 12-inch girth. About 8 cents per foot run.

62-foot run. Painting in four oils, window-sills about 12-inch girth. Price about 8 or 9 cents.

Painting in approved tints wood and stone chimney pieces, four coats. If of ordinary kind, the cost may be put at about 75 cents to $1.00 each. Ditto ditto, extra coat and flatting. Add, say, 30 cents each.

30 yards super. Painting four times in oil, including knotting and stopping and flatting.

Say for four-coat work on wood	$0.25
For flatting add	.08

In some price-books this would be put at 40 cents per yard.

26 yards super. Ditto ditto finished in party colors. Add 5 cents to the above.

5 yards super. Ditto finished in shades of Indian red. This is rather a dear color, and may be priced at 8 to 10 cents in addition.

60-foot run. Paint in three oils, reveals 4½ inches wide. Add about 6 cents per foot.

58-foot run. Ditto three and flatting to skirting not more than 10 inches wide. About 7 cents per foot.

10 yards super. Painting in three oils, enriched cornices and flatting. Price about 75 cents per yard, and add 20 cents per yard for flatting.

No. 12. Sash frames not exceeding 24 feet super., four oils. These may be priced at about 80 to 90 cents each.

No. 4. Ditto large size ditto. Add 25 cents to each.

No. 12. Dozen sash squares, about 2 feet super. each. Worth about 55 cents per dozen.

No. 4. Dozen ditto large. About 80 cents per dozen.

72-foot run. Painting base, four oils. These would be about 7 cents per foot.

72-foot run. Ditto finished in grayish-green. Add 1 cent per foot.

32-foot run. Ditto narrow base, four oils. About 7 cents per foot.

GRAINING AND VARNISHING

18-foot run. French-polishing handrail. Worth about 20 cents per foot.

50 yards super. Varnishing doors and framing, two coats copal varnish. Price at 20 cents per yard super.

45 yards super. Painting in four oils, doors finished in buff and gray of approved tints.

Price in common colors, four coats, including knotting and stopping, per yard.................$0.25
Finishing in fawn tints, per yard................ .06

62 yards super. Graining extra in oak and twice varnishing. This may be priced at 50 cents per yard for best work, and for twice in copal 30 cents.

105 yards super. Graining wainscot and twice varnishing. Extra over common.

Graining cost per yard.........................$0.30
Copal varnishing, two coats.................... .20

320 yards super. Varnishing matchboard partitions, etc., in two coats copal varnish, and sizing wood.

Sizing wood, say..............................$0.10
Twice in copal, say.......................... .20

32 feet super. Painting carved pediments and trusses four coats in oil, finished in two tints to be approved.

Say cost of four-coat work....................$0.12
Picking out in two tints, per foot............... .08

If very elaborate, the cost would be more, according to color selected.

32 yards super. Painting skylights each side four coats. The price would be about 28 cents each side.

12 yards super. Oak combed and shadowed and varnished. This may be for some special doors, and may be priced at 68 cents per yard.

If there are more yards in the work than named in the foregoing, then a reduction of from 3 to 5 per cent may be made. If there is a less number of yards, then an additional price of from 3 to 5 per cent may be added.

THE PLASTERER AND PAINTER

In estimating for plastering, or for painting also, (1) the description of all materials and work should be kept separate. (2) Plastering on walls to be measured from the floor upwards, or from the point where each description of work commences. (3) Where cornices are lathed on brackets, measure ceiling and walls to the edge of the brackets only. (4) Where cornices are not bracketed, measure the ceiling full size of room, and the walls up to ceiling; all in super. yards. (5) Deduct all openings 100 square feet and over; deduct materials and add labor (hollows) for net sizes of doors, windows, fireplaces, and other openings under 100 feet super. (6) Where ceilings are paneled and coffered, or covered, girth round all portions that are lathed, keeping circular work separate. (7) Ceilings plastered between spars, etc., to be measured across the spars and purlins, and even then kept separate, and described as such. (8) All work run with a mould to be measured lineal on the wall, and the girth given, as cornices, rustics, strings, architraves, soffits, quirks, etc.; count all miters with the girth of mould they belong to; count miters in paneled work. (9) All cornices, etc., lathed on brackets, to be kept separate, and described as such. (10) All cast work to be counted, except running enrichments. (11) Enriched members to be measured lineal, with girth. (12) Modeling of enrichments to be, if special, so stated, and the models to be the property of the designer. (13) Ceilings or walls covered with panels, formed by small moulds, to be measured super., with illustration or drawing, for "extra price over plain work"; larger paneling or special decorative features to be measured in detail. (14) Angles to pilasters, etc., if specially formed, lineal and extra to plastering. (15) Door and window frames, bedding and pointing, to be counted, and state material to be used; also flushing to inside of frames after fixing, or behind casings, window backs, or other work to be given. (16) Making goods generally, and after plumber, gas-fitter, bell-hanger, etc., and chimney pieces, as in item, stating numbers. (17) Coloring and white-washing walls, etc., to be in super. yards, measuring over all openings under 100 super. feet; if the work has to be pointed by the plasterer, state so. (18) Painting to include stopping and knotting, and to be given in square yards. Priming to be separate, if on work painted before being fixed. Painting to be girthed round all exposed surfaces, except as below. (19) Balusters, if ordinary square, and girds, gates, and other metal work painted on both sides, with bards about 5 to 6 inches apart, to be measured one surface only; if closer or slightly ornamental, 1½ surfaces, and for very close or very ornamental work, 2 to 2½ surfaces. (20) Windows to be measured each surface over full size of opening for painting frame and sheets, or else the frames counted, and the sheets, if large squares, counted; but if in small squares (as old-fashioned crown glazing), then count the squares instead of the sheet. (21) Fancy or ornamental painting to be measured in detail, with lengths of mouldings picked out, gilt, etc. All work in parti-colors to be kept separate from plain work.

The cost of internal plastering largely depends on the number of coats; the second or floating coat involves four processes: running the screeds, filling-in, scouring with a hand-float, and "keying" the surface for the finishing coat. This coat costs about a ½c. more than the two coats and set. The third or finishing coat also entails extra care and trouble. It involves laying, scouring, troweling, etc., and it requires "fine stuff," consisting of pure lime, slaked, saturated till semi-fluid sand. If "gauged" with plaster of Paris in the proportion of three or four to one, the work dries quicker. This is also used for cornices and enrichments. Gauging with plaster costs about 8 cents each coat per yard extra, and therefore adds materially to the cost. The cements known as Keene's and Parian have quick-setting properties, and give a hard, non-porous surface; they are laid in two coats, the first of cement and sand about ½ inch thick, and the finishing coat of neat cement. This kind of cement finish is used for angles and arrises, often on Portland cement grounds, also for mouldings, girder-casings, soffits, skirtings, and other decorative features. Compared with ordinary three-coat work, it costs about one and a half times as much. Some authorities give 70 cents per yard on brick, and others 20 on lath, including profits, and on Portland cement grounds.

There are several patent fibrous plasters used on canvas, wood, and metal for ceilings and decorations, that are advertised. These vary in price about 28 to 40 cents per yard. The estimator can obtain prices for any selected ceiling, wall filling, or decoration.

In estimating items of plasterer's work, care is necessary in ascertaining the quantities, and whether for "narrow widths," or for circular work. If for narrow widths, an extra price is necessary, being for labor, which would come to about 6 cents per foot super. more, or 7 cents if in plaster of Paris. The quantity should also determine the price; for large quantities the labor might be priced at 1 cent less. Keene's fine quality cement takes a fine polish, and is used for internal decorations, panels, columns; on brick walls it should be applied on a rendering coat of Portland cement. Parian cement is used as a stucco, and is valuable on new-built walls, as it can be papered or painted very soon afterwards; 4 bushels of Parian to 4 of clean washed sharp sand will cover 10 super. yards ½ inch thick. The price is about the same as Keene's cement.

Rake out joints of old brick work to form "key" for plaster.

This may be done in brick work for 3 to 5 cents per foot super., say 32 cents per yard, and the price depends much on the hardness of the mortar to be raked out. Raking out cement joint would be about 6 cents per foot.

Dubbing out 1 inch thick in tiles and cement to fill hollow in wall. This may be taken at from 9 to 13 cents per foot super., according to the kind of wall, and whether a scaffold is necessary.

Render, float, and finish in troweled stucco for paint. May be put down at 35 cents per yard on brick. Add for last coat finished troweled stucco for paint 13 cents per yard. Troweled stucco on lath would cost about 9 cents per yard more.

Lath, plaster, and set, finished troweled stucco in narrow widths. This would come to about 9 cents per foot super.

Ditto sloping ceiling in panels between ribs.

Say ordinary work	$0.50
Extra for lathing, say	.18
Add for setting coat between ribs	.05
Per yard	$0.73

Moulded cornice, 15-inch girth. Price this as before, say, 28 cents per foot super.

Cornice, 5-inch girth. Worth about 14 cents per foot run.

Miters to ditto. Each, say, 14 cents.

Ditto 9-inch girth. Worth about 30 cents.

Miters to ditto. 28 cents each.

Enrichments 7-inch girth to detail; at 5 cents for each inch girth per foot, would come to 30 cents per foot.

Render, float, and set walls, gauged with equal quantities of lime and cement. Add 13 cents per yard to former price, say, 34 cents.

Ditto in narrow widths. Price at 50 cents.

If circular. About 50 per cent more than the straight.

Hacking face of old walls to form key for plaster. This is labor only, and may be put down at 5 cents per yard.

Ditto and raking out mortar joints. Add another 5 cents per yard.

Rendering chimney backs. Worth about 25 cents each.

Plaster plain face on brick in narrow width. If this is for lime and hair finished with setting stuff, it may be priced at 50 per cent more than for ordinary plastering; the difference is entirely for labor.

Plain face in Portland cement for skirting 10 inches high with sunk bead on top. Worth for plain face about 5 cents per foot.

Worth for plain face about 6c per foot	$0.06
Bead, per foot	.10
	$0.16

PLASTER CORNICES AND ENRICHMENTS

Moulded cornice, as per detail, on lath. This item may be priced the same as previous item, adding lathing, say, 4 cents.

Papier-maché center flowers to drawing-room and dining-room, about 3-foot diameter, according to design. It is not easy to price this item without seeing the design, as they vary according to the degree of enrichment. For plain designs we may price them at 20 cents per inch diameter. For elaborate designs, 30 to 60 cents per inch would not be too much. Get list of prices.

Plaster center flowers, 18-inch diameter. These are worth about $3.50 each.

Ceiling decorations, as per design. No special decoration is described; if plain, the cost would be about 10 and 14 cents per foot, and fixing, say, another 12 cents.

Cornices to ditto to design. Price from 20 cents, for fixing add 20 cents per foot.

Frieze. About 30 cents per foot, including fixing.

KEENE'S CEMENT

Keene's cement, coarse quality, on brick walls, on rendering of Portland cement. Troweled on brick, at 70 cents per yard. This includes profit.

Ditto on single-lath partitions. Price at 78 cents per yard.

Ditto circular ditto. Add 14 cents per yard.

Pilasters and architraves ditto. This item depends on detail; 14 cents per foot for plain work would do.

Skirting 9 inches high and moulding 3-inch girth. About 20 cents per foot.

Miters to ditto. About the same price each.

Enrichment, 12-inch girth. About 12 cents for every inch girth per foot run.

Moulded cornice, 15-inch girth. Price at 50 cents.

Angle 6-inch girth, and arris in Keene's cement. Worth about 10 cents.

Staff bead 2-inch girth and quirks. About 14 cents per foot.

Moulding on ditto 4-inch girth. 16 cents per foot run.

Keene's fine quality cement, on Portland cement grounds, polished face, in narrow widths. This is priced at 75 cents, including profit, per foot.

Ditto polished, plain face, on lath partition. This may be put at about the same.

Ditto to pilasters on brick. More labor is necessary in troweling and floating the surface of diminished pilasters, and the cost would be about 75 cents per foot.

Ditto to columns. Add 14 cents to last.

Ditto in No. 2 spherical heads of alcoves 6 feet wide each. The price for these would be about the same per foot super. There would be about 28 square feet in each head.

18 feet super. Moulding to ditto polished. The price for these is about 75 cents to $1.10 per foot; for circular work, another 20 cents may be added.

Arrises. Put at 6 cents per foot.

Moulded cornice round saloon bracketed with two enrichments, per detail. (See Fig. 16.) This cornice is run on lath, bracketed out, and the items may be put down thus:

1½-inch pine brackets and plugging, per foot	$0.14
Moulding per foot super., say	.30
Two enrichments	.35
Add for lathing	.03
Per foot super	$0.82

Miters to ditto.

As these entail extra labor, they may be put down equal to 1-foot run of cornice, which is equal to nearly 2 feet super., say, 80 cents each.

I show several examples in decorative plastering in Figs. 11, 12, 13, 14, 15, and 16, which will give some idea of the character of work estimated on in the foregoing analysis, and aid the estimator in working out his figures.

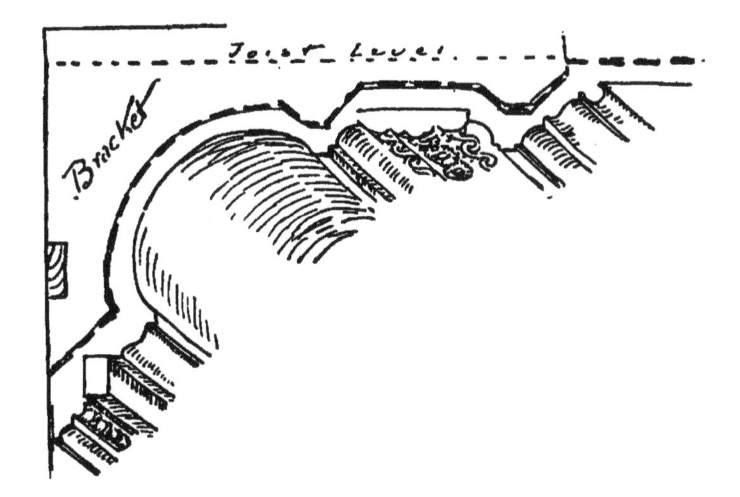

Fig. 11.

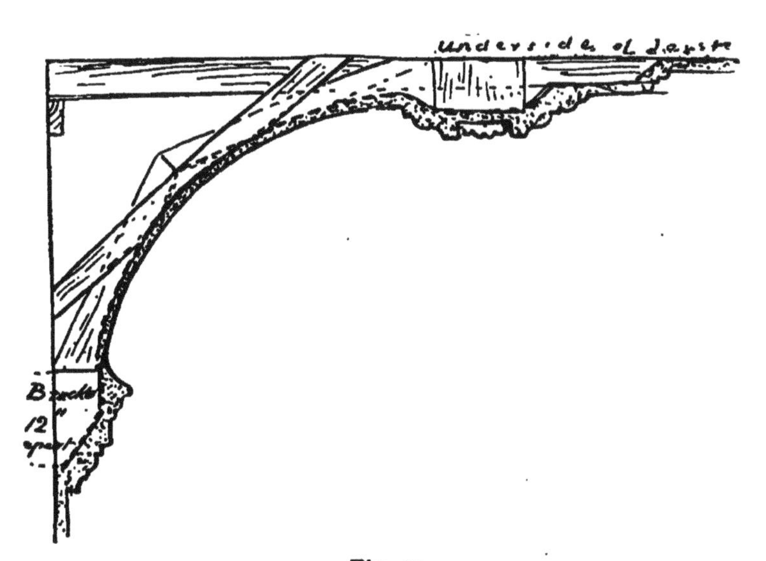

Fig. 12.

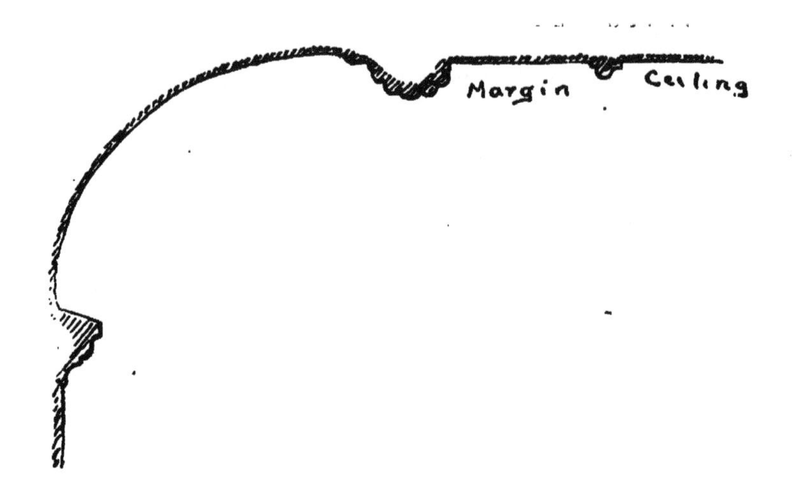

Fig. 13.

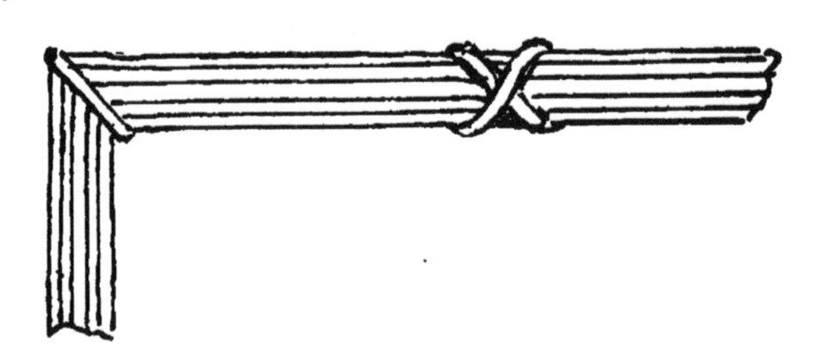

Fig. 14.

Fig. 15.

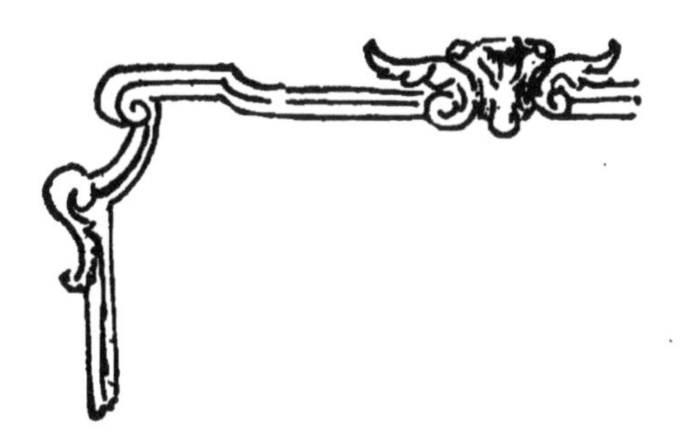

Fig. 16.

EXTERNAL PLASTERING, PORTLAND CEMENT

Work in Portland cement is costly, as both the material and labor are heavy. Portland cement is mainly used in external work. Plastering will take 3 of sand to 1 of cement, except in reservoir or hydraulic work, when it is used nearly neat. It is generally sold by the bag of 100 pounds in place of the bushel, and its price is variable according to the quality and locality. Price lists should always be kept on hand. A gritted face is better for Portland cement finish than troweled work, as the latter is apt to blister. One bushel of cement and two of sand will cover 4½ yards ¾ inch thick; one of cement to three of sand will cover about 6½ yards of that thickness. So that practically we may say that one-sixth of a bushel of cement will cover one yard at the above thickness, worth about 12 cents; and three bushels of sand will cost 60 cents, one-sixth of which will be also about 10 cents; add to which labor and profit, about 37 cents, making about 60 cents per yard. The items should clearly state whether the cement is "rendered" or "plain face," or "render and float," or "troweled," if on brick or lath. Mouldings, grooves, arrises are taken per foot run if not exceeding 12-inch girth; if above this, by foot super.; and all narrow widths, like panels, rustics, etc., should be stated, as the labor is greater.

OUTSIDE PLASTERING

Troweled rough stucco, with blue lias lime and good sand, and jointed. This is worth about 50 to 55 cents per yard.

Moulded cornice ditto, 2-foot 6-inch girth. May be priced at 28 cents per foot super.

Portland cement, weathering, dubbing, etc. The labor would be about 7 cents, the material about the same, say, 14 cents per foot.

78 yards super. Portland cement, plain face. If 1 of cement to 3 of sand, the cost would be about 55 cents.

Ditto in narrow widths. About 10 cents per foot super.

Ditto 1¼-inch thick rustics in narrow widths. This may be put at the same, as these rustics are in narrow widths between grooves. This would be for the plain face; vermiculated work costs more.

Rough-cast made with clean washed sand and shingle and good lime in proper proportions.

Say, materials per yard super	$0.08
Labor	.30
	$0.38

Rough-cast on brick, with washed sand and shingle and Portland cement. Add 14 cents to former item, for a yard of cement ¾ inch thick will take about one-sixth of a bushel, say, 12 cents; add for extra labor 3 cents.

PORTLAND CEMENT WORK

460 yards super. Portland cement, plain face on brick, floated; 1 of cement to 3 of sand.

Say, cost of rendering (cement and sand)	$0.25
Labor to ditto	.14
Floating, add 14 cents	.14
	$0.53
Profit, add	.03
	$0.56

The price-books put down for this work 65 cents per yard, which is high for a large quantity.

Ditto in narrow widths. Price at about 10 cents per foot super., or about 90 cents per yard super.

Ditto plain face on brick-jointed. This may be put down at above price, or

Plain face	$0.55
Add for jointing	.03
	$0.58

and 80 cents for circular.

Plain face ditto as plinth. This may be put down at 8 cents per yard more than last, or, say, 90 cents.

Moulding, etc., ditto, 9-inch girth. Price at 35 cents per foot.

Plain cement face to pilasters, etc. Worth about 10 cents per foot.

4½-inch reveals to windows and arris. May be priced at 8 or 10 cents per foot.

Vermiculated work according to sample for quoins. Worth about 55 to 70 cents per foot. This will be extra on the plain face before taken.

Quoins 18 inches long, 12-inch return, 12 inches in height, including dubbing out and projecting 1 inch from face. These are plain, and would cost about 15 cents each.

Returned and mitered ends to moulding. The cost of these may be put at the price of 1-foot run of moulding; a sketch should be given. Put at about the same.

Portland cement plain face. This is worth about 55 cents per yard (see previous items).

1¼ inch thick in rustics. This is chiefly for labor, and may be worth from 55 to 85 cents or more per foot, but the price depends on the class of work.

Moulded grooves to rustics, as per detail. This will be worth from 8 to 14 cents per foot, according to girth.

Miters to ditto. Worth about 10 cents each.

Rustic grooves, V-shaped. These are simpler than moulded, and the labor is less, say, 5 to 8 cents per foot run.

Miters to ditto. Put these at 3 cents per inch girth.

Portland cement cornice, per detail. If the moulding is plain, the work may be done for 38 cents per foot.

Ditto in short lengths. Add 10 cents per foot.

Miters. Say, 42 cents each.

Floating beds on concrete for tile paving.

Say, cost of cement and sand, etc.	$0.20
Labor, say	.25
	$0.45

Floating beds on concrete for wood-block paving. This may be priced the same.

Portland cement laid as paving, 2 inches thick. The cost would be about 5 cents for materials, 8 cents for labor, or 13 cents.

Selenitic cement grounds for Keene's cement. This cement forms a good ground, and can be worked to a smooth face. Obtain price and instructions from the manufacturers or dealers.

PORTLAND CEMENT

Angle 8-inch girth, and arris. If circular add 50 per cent to previous prices.

Splayed angle, 8-inch girth, and arris. This may be priced at the same as the other angle.

6 inches by ½ inch square skirting and dubbing out. 6 inches of plain face in narrow widths at 50 cents; add 3 cents for arris and narrow return, and add 3 cents for dubbing; 3½ inches in all.

Miters. Price these at 1-foot run of the skirting.

Stops. Price these at one-half the last item.

Moulded skirting 9 inches high and 1 inch projection, and dubbing out. Say, 4 cents for plain face, 4 cents dubbing and 5 cents for small moulding.

Ditto, ditto, raking, and ramped over steps and risers. Price this at 50 per cent on last price.

Both these prices are rather high.

Internal miters. These are worth 1-foot run of the straight skirting; but the above prices are sufficient to cover the cost of miters and other extra labors.

External miters. These are of the same value.

Stops. These are worth half a miter.

TILING

The cost of tiles and tiling can only be accurately ascertained by first getting price of tiles and cost of labor in laying them. These are laid in different ways; a door boarding, on cement, or on laths or battens. The latter method is that generally employed. Tiles, in shape, are of two main classes; those which, like pantiles, interlock, and those which, like common plain tiles, are nearly flat, and are laid on the same principle as slates. In the former class innumerable forms have been patented, but few of them get into general use, chiefly owing to difficulties of replacing when broken, and the trouble of fitting them to irregularly-shaped roofs. Plain or crown tiles are such as have a rectangular form and plane surface. A custom is supposed to regulate size, but they are generally 10½ inches long, 6½ inches broad, and ½ inch thick, with two holes in them, through which oak pins are inserted to hang upon the laths. Sometimes cast-iron pegs are used instead, or frequently extra large flat-headed wrought nails, made of pure zinc or zinc and copper, which have the advantage of allowing a tile to be replaced from the inside of the roof by lifting up the others to place in the tile and drop in the nails in a few seconds. Sometimes, also, tiles have projecting nibs cast on in lieu of pegs, or they may be both holed and nibbed, so that if the nib is broken off, the tile may be nailed. In use, one tile laps over another, and that part which then appears uncovered is called the gauge of the tiling, likewise known as the fade or weather. Many tilers have a practice, when plain tiles are set in mortar, not to peg more than one hole in ten, or sometimes only every third or tenth course is nailed. This is bad, as with the decay of the mortar, the tile will slip down. For walls, battens, nailed or plugged to walls, are the best mode of fixing for vertical tile-hanging, the top of each tile being bedded in cement mortar, and the bottom double course bedded and pointed in cement on a tilting fillet.

In dealing with tiles as a roof covering, the first thing to be sure of is that the tile selected is capable of excluding all damp, and will withstand the disintegrating influence of the weather.

Pantiles are the commonest class of tiles, and are

very cheap. They hold moisture a long time, and require extra strong roof timbers. They are best laid to a slope of about 24 degrees, and are mostly used for covering sheds, barns, and buildings which do not require a plastered ceiling.

Plain tiles are smaller than pantiles, but being laid with more lap, are heavier per square. They can be laid to any slope from 25 to 60 degrees.

Fancy roofing tiles are similar in many respects to plain tiles and are much used for external walls of half-timbered houses in some countries.

Roofing tiles are subject to the same defects as terra-cotta, viz., if they are burnt thoroughly many of them twist and warp and will be found to be untrue, and if they are not burnt very hard they are liable to decay.

A good tile should be well tempered, of good color, free from stones, carefully trimmed, should give forth a clear ringing sound, and take its weathering quickly.

The characteristics of a good roofing tile are density, toughness, and incipient vitrification, the last named quality producing, to some extent, that pleasing tint familiarly known as "bloom," one of the peculiarities of some makes of tiles.

Among the best are Jersey tiles, the color of which varies from pale strawberry red to dark brindle (a deep reddish brown), or even to blue, through an almost infinite gradation of color, so that almost any color can be obtained. They get their weathering quickly, and are not porous.

Sizes of tiles. Plain tiles, 10½ inches by 6½ inches by ½ inch, and weigh about 2½ pounds each, and 11 inches by 7 inches by ⅝ inch, and weigh about 3 pounds each. Pantiles are 13½ inches by 9½ inches by ½ inch, and weigh about 5¼ pounds each.

A square of roofing requires 800 plain tiles laid to a 3-inch gauge, 700 tiles laid to a 3½-inch gauge, and 600 tiles laid to a 4-inch gauge; and 180 pantiles laid to 10-inch gauge, 164 pantiles laid to 11-inch gauge, and 150 pantiles laid to 12-inch gauge.

One square of plain tiles weighs about 15 cwt., and one square of pantilès about 8 cwt.

Spruce tiling laths or battens are 1¼ inches by ¾ inch to 2 inches by 1 inch, and oak tiling laths 1¼ by ½ inch.

100 plain tile laths 5 ft. long....................1 bundle
12 pantile laths, 10 ft. long....................1 bundle

One square of tiling requires 1 bundle of laths, 12 hundred of nails, 1 peck of tile pins, and three hods of mortar.

One square pantiling requires 1 bundle of laths and 1½ hundred of nails.

To ascertain the comparative merits of tiles, as to their weathering properties, there is no better test than the amount of water they will absorb.

Most roofing tiles are slightly absorbent, except in the case of highly-fired brindled and blue tiles, and for this reason old tiles have, in a few isolated instances, attained to a higher market value than new tiles, as by age and atmospheric deposit they have acquired an artificial surface coating and lost the property of absorption, at least on their outer exposed surface.

Tiles of a bright red, or an earthy red, color should be viewed with suspicion and avoided. They are invariably absorbent, and will not weather well. Tiles may be obtained of almost any color.

Well-formed roofing tiles are straight in their width and hollow in their length, that the tails of each course may lie close and tight on the backs of the under course.

Straight tiles will not clear themselves one over the other, and should therefore be rejected. Where pointing is necessary, it is customary in good work to grind down some of the broken tiles, to mix with the Portland cement as a substitute for sand, that the finished pointing may approximate in color to the general tone of the roof covering.

The gauge of tiling is the distance from head to tail minus the lap divided by 2; thus a 10½-inch tile laid to a 2½-inch lap will only expose 4 inches of its length to the view when the work is completed.

FIXING

Old-fashioned tiles have no nibs or stubs for hanging, and must therefore be kept *in situ* by means of two wooden pins or nails.

When tiles are bedded in lime and hair mortar the tops only should be bedded, the mortar extending, say, not more than 3 inches below the head of the tile.

When a roof is close-boarded (and sometimes felted) there is no need for bedding, though of course a covering of bedded tiles is less liable to breakage when a man is climbing about a roof than would be one of unbedded tiles.

In tiling roofs it is well to cover them with roof boarding and felting before laying the laths and tiles. This should always be done in good work. Occasionally architects are compelled, owing to the cost of work, to eliminate the felting, or covering with 2-ply paper, and lay the tiles on the boarding; but this practice is unwise, as experience shows, and the boards alone do not stop draughts.

Secret gutters should only be used in positions

where they will always clear themselves, especially if the dwelling be surrounded by trees.

Tiles, and tile-and-a-half, should be worked against all secret and other gutters, where practicable, alternately on each course.

Tiles overhanging secret gutters should not be bedded on the lead, nor should their edges be pointed, otherwise rain may be drawn into the roof.

The lead welt should stand its own thickness above the backs of the battens, forming a tilt for the tiles, so as to throw the water away from their edge on to the main body of the roof.

Ridge tiles should be of such a section as to admit of being pressed or made in one piece. Where an ornamental cresting is required, it should be made as a separate piece entirely detached from the ridge tile proper, the latter being made with a groove to receive the cresting.

The cresting should not be stuck on the ridge tiles by means of semi-liquid clay while they are in the clay state. Such work is more or less defective and unsatisfactory in the end.

Pantiles should be laid on laths and a good bed of hair plaster, in order to secure them to the roof.

Tiles hung against vertical walls are treated precisely as are those on roofs.

HIPS, VALLEYS, AND RIDGES

In a tiled roof valley and hip tiles should be used in preference to lead gutters, secret or otherwise, bedding the valley tiles at their heads to keep them *in situ* and steady while laying the plain tiles.

Hip and valley tiles should be purpose made, with proper regard to their enclosed angle or pitch.

Where a minor roof runs in at right angles to a greater or main roof, intersecting it at a point below the main ridge, it is desirable to use a piece of 4-pound lead dressed to the shape of the minor ridge and the slope of the main roof, and called a saddle-piece. This prevents the possibility of rain getting in at the junction of the roofs.

The simplest form of ridge tile is that consisting of the two wings terminating in a roll at their angle of intersection.

Another good form of ridge tile is that of a plain vertical blade rising from the angle of intersection of the wings, and with the square angle at each end of the blade cut off at an angle of 45 degrees, and which can be pressed in one piece by a simple operation.

Ridge tiles should be well soaked before use, bedded in gauged lime-and-hair, and their vertical joints drawn up solid with cement, not simply pointed after they are fixed.

When the roof is enclosed on the under side, it is customary to bed in lime-and-hair the eaves courses only, for the sake of steadiness in the fixing.

As before stated, the cost of a tile roof will vary much according to locality and quality of materials used. The average cost per square, however, will be about $16.00 for the best tiling and about $12.00 for the more common kinds. While these figures are not correct, they may act as a sort of guide to the estimator when figuring on tile roofs. In all cases, however, wherever possible, I advise that the local prices be obtained and that at least 10 per cent be added to these prices, unless the work is executed in a large city where prices are more constant than in country places; then only the usual percentage of profit be added.

So little tiling is done in this country (more the pity) that expert tilers are scarce and wages high and varied, so that nothing can be given definitely regarding the cost of this work.

In measuring for tiling, take the whole superficial area, and allow extra for eaves next parapets, 4 inches; dripping eaves, 6 inches; all hips and cuttings, 3 inches, and for valleys, 12 inches.

For pantiling, also take the whole superficial area, and at hips, take the length of the hip-rafter by 12 inches for cutting and waste, to be added to the superficial area; take the run of hip and ridges, and of mortar or cement filleting, and the plain tile heading.

Take in all cases the number of hip hooks and T nails to be painted in oil.

Secure gauge of the tile, the quantity and description of the laths and nails used; also if laid dry or pointed outside or inside with mortar or cement, and charge up accordingly; get exact cost of one square according to data given in the foregoing, and then find number of squares to be tiled, and multiply the number of squares by the cost of one square estimated upon.

THE SLATER

The great similarity which runs through the specifications for slaters' work, no matter by whom drawn, or for what class of work it is intended to apply them, is a mistake, as it often leads to bad results. The most suitable slate for the particular work in hand should be carefully selected.

The architect should consider the pitch it is intended to give the roof, the length of span, and also whether mitered or close-cut hips are to form any portion of the roof.

If the hips are to be mitered, the angle should not be less than 45 degrees, otherwise very large slates must be used at the hip, which looks unsightly, and on no account should small pieces be allowed.

Soakers should invariably be used where soft slates are laid, as flushing or bad work of any sort stains the slates and produces a bad effect.

In exposed situations, where snow may be driven over the lap, it is better that the roof should be boarded and felted. If battens are used instead, vertical ones are less liable to cause a collections of snow at certain points, and apparent leakage when that occurs.

When snow may slide off main roof on to any glass below, wire guards should be fixed along the eaves to check it. Open batten show-guttering should be provided to all V and parapet gutters to allow snow-water to get away.

Mitered hips and valleys with 4-pound lead soakers under slates make the neatest finish to slated roofs, and, if properly secured, the most satisfactory. In order to make a neat finish the roofs should be 45 degrees pitch and the slates used in such cases should be small, say 16 × 8; the slater has then the choice of such sizes as 16 × 9, 10, and 12 to work up the hip with. It is impossible to obtain wider slates, and this often induces the slater to lay the slate lengthways to save the introduction of small pieces; the sides of the roofs forming the miter should be of the same pitch.

If additional precaution is deemed necessary, small [illegible]ls may be screwed down to the hip rafter, over the [illegible]ering; this is rather unsightly and not recommended if soakers are used.

Slates should be nailed with copper nails, which are practically imperishable. The life of a zinc nail rarely exceeds twenty years, and iron still less.

In soft and rag slating the nails should be very stout, and the length 2 inches, 1¾ inches, and 1½ inches; few of the latter, if any, should be used, say on the last three or four courses only; the strength should be 90, 110, and 130 to the pound respectively.

In regular-sized work such stout nails are inadvisable, as the heads are large and will not recess as readily into the slate, and the top of the head must be flush with the surface of the slate, or anything pressing on that particular part will damage the slate above; 1½-inch and 1¼ inch nails are recommended, 180 and 250 to the pound respectively.

Gauge.—The gauge of slating is the part left exposed, viz., deduct the lap from the total length of slate and half the remainder, thus,

$$20 - 3 = 17 = 8\frac{1}{2}$$

Lap for soft and rag slating should never be less than 3 inches. For regular-sized slating, 4 inches to 2½ inches, according to pitch of roof.

Repairs to roofs should be done by an experienced slater and straps prohibited; the lead or zinc strap is a ready way, but raises the tail of the slate up, and is turned back by snow slipping down and slate slips with it.

Slating on unplaned boards is preferable to that on battens, because it is more waterproof and prevents the ingress of driving snow. The cost of good quality ⅞-inch rough board is about $2.00 per square as compared with 50 cents for 2 × 4 inch slate battens, and the labor of laying and quantity of nails equal in each case.

In superior work heavy felt (inodorous or otherwise)

is inserted between the covering boards, and the slates or battens may be added above the felt to render the building more proof against sun heat.

Bedding and pointing on under side is not recommended unless the roofs are well ventilated; the heat of the house will condense on the under side of slate and quickly rot the wood work, and, in course of time, the slate also. Experience shows that a rough slate will keep out driving rain better than a smooth one, if well laid; the reason for this is that there is a considerable quantity of air between the surface of rough slate and practically no suction; also the thick edge of the slate breaks up the force of the wind on the surface of the slates.

In church roofs, where the pitch is very sharp, small slates are recommended, from 14 x 8 to 18 x 9, according to pitch; as the pitch decreases the slate should be wider.

For roofs of warehouses, where much depends on the work being perfectly water-tight, "tin" slates are recommended; they are about $\frac{3}{8}$ inch thick, and are large and laid in diminishing courses, the gauge being about 15 inches at the eaves and 10 inches at the ridge.

These slates are scarce, and architects should insist on the order being placed when the contract is signed, to ensure delivery in time.

If it is thought advisable to use the above-mentioned slates, sizes such as 16 x 12, 18 x 12, 20 x 22, 22 x 12, and 24 x 12 are suggested, the size varying according to pitch.

For curb and mansard roofs, slates larger than 16 x 9 should never be used, the whole weight being thrown on the nail in such cases. The appearance of small slates is also far better on such roofs.

With span of 25 feet to 40 feet, which entail a deep rafter and flat pitch, it is a wise precaution to vary the lap, giving extra at eaves and for a third of the way up the slope; in such roofs the slates should not be less than 11 inches wide, the extra width being a safeguard against side leakage through the nail holes.

Merchants are only able to obtain a proportion of sizes yielded by the rock, consequently it is sometimes impossible to fulfill the general specification of 20 x 10 Countess slate, as the quarries will not sacrifice the rock to make the full demand of 20 x 10 if the block will make 20 x 12; if 20 x 10 were insisted on, in that case it would entail an extra cost of about $3.50 per square over other sizes.

Actual size has little to do with the quality of the work, the lap is the principal factor, and the result in 16 x 10 or 20 x 10 is exactly the same. The best all-round size is probably 16 x 10.

Single samples are very unsatisfactory means of judging of the quality of the bulk; at least six should be demanded, showing medium and thinnest. Where possible, the inspection of the bulk should be made.

A good slate is hard and tough, will give a sharp metallic ring when struck with the knuckles, does not split under the slater's ax, is easily holed without fracture, not tender or friable at the edges, and should contain no white iron pyrites (marcasite).

A bad slate feels smooth and greasy to the touch, absorbs moisture if stood in water, splits while being holed or trimmed at the head, breaks when pressed upon, emits a clayey odor when breathed upon, and is liable to premature decay.

Slate ridge rolls and wings should be fastened with

brass or copper screws, and bedded and pointed in lead cement, one-third lead and two-thirds best oil putty. Iron screws should not be permitted, they oxidize and burst the rolls. If wings more than ¼ inch thick are used the upper edges must be beveled.

Half or checker slating is sometimes employed for farm buildings or where special ventilation or cheap covering is required. The saving by this method is in the quantity of slates and nails used; the battens or boards remaining the same. In place of the slates being butted close to one another, they are spaced laterally in such a manner as to just cover the joint between the slates in the course below. This slating, known also as open slating, is well adapted for use in farm buildings, covered yards, etc., as by its construction it affords a certain amount of ventilation.

In laying slate there is always an element of risk of breakage that must be accounted for, and, as all roofs must be left in good order and perfectly water-tight, an allowance of about 25 cents per square must be made above all other provisions. It is very necessary to go carefully over the slating and see that the slater who does the work makes good any deficient or broken slates before he leaves it; and beyond that there is the risk of breakages from other workmen, for some men must go on the roof after, although as much as possible this should be avoided.

Cutting round small ventilators, V-shaped on plain, and 12 inches by 12 inches.—If the ventilator itself measures 12 by 12, the flashing round it will, of course, exceed the dimension and the slate will not run close up; giving another foot run of cutting, the slate would have to be *tilted* against the ventilator to throw the water off, or a secret gutter formed. The eaves cutting price at 10 inches of the slating, the plain cutting at 6 inches of it.

Cutting round 3-inch lees pipe (ventilators from soil-pipes) and making good.—These are at $1.00 each, including profit.

The following prices are given herewith as being approximately correct, being taken from the price list of the Slatington Slate Co., Slatington, Pa., but I would advise estimators to get other price lists, down to the latest date, as the prices are continually changing.

BANGOR NO. I BLACK ROOFING SLATE

Sizes	No. 1 Price per Sqr. F. O. B. Quarries	No. 1 Ribbon Price per Sqr. F. O. B. Quarries	Sizes	No. 1 Price per Sqr. F. O. B. Quarries	No. 1 Ribbon Price per Sqr. F. O. B. Quarries
24 x 14	$3.50	$3.10	16 x 12	$3.85	....
24 x 12	3.50	3.10	16 x 10	4.25	3.50
22 x 12	3.85	3.25	16 x 9	4.50	....
22 x 11	3.85	3.25	16 x 8	4.50	3.50
20 x 12	3.85	3.25	14 x 10	3.85	3.50
20 x 11	4.25	....	14 x 8	4.25	3.50
20 x 10	4.25	3.35	14 x 7	4.00	....
18 x 12	3.85	....	12 x 8	3.75	....
18 x 10	4.25	3.35	12 x 7	3.50	....
18 x 9	4.40	3.50	12 x 6	3.50	....

Add 15 per cent to above prices.

BROWNVILLE MAINE SLATE

No more beautiful slate is quarried in the world than the Brownville. It is very uniform in thickness and of smooth surface, and when laid on the roof presents a surface equal to polished steel. For costly private residences, churches, and public edifices, it has no superior.

Sizes	Price per Square F. O. B. Quarries No. 1	No. 2
24 x 14	$6.20	None.
24 x 12	6.20	$4.95
22 x 14	6.10	None.
22 x 12, 22 x 11	6.30	4.95
20 x 14	6.50	None.
20 x 12, 20 x 11, 20 x 10	6.70	4.95
18 x 14	6.45	None.
18 x 12, 18 x 11	6.75	4.75
18 x 10, 18 x 9	7.20	5.20
16 x 12, 16 x 11	7.00	5.00
16 x 10, 16 x 9, 16 x 8	7.20	5.20
14 x 12, 14 x 10, 14 x 9, 14 x 8	6.45	4.60
14 x 7	6.25	4.50
12 x 10, 12 x 9, 12 x 8, 12 x 7	6.00	4.10
12 x 6, 11 x 8, 11 x 7	5.20	3.45
10 x 8	5.00	3.45
9 x 7	4.00	None.

Add 15 per cent to above prices.

GREEN, PURPLE AND RED ROOFING SLATE

For ornamental roofs these colors are in steady demand. They are also used for entire roofs in many instances.

Sizes	Unfading Green. Price per Square F. O. B. Quarries	Purple. Price per Square F. O. B. Quarries	Red. Price per Square F. O. B. Quarries
24 x 14, 24 x 12	$3.50	$4.00	.....
22 x 14, 22 x 12	3.50	4.00	.....
20 x 14, 18 x 12, 16 x 12	3.50	4.00	.....
22 x 11, 20 x 12, 20 x 11	3.75	4.25	.....
18 x 11	3.75	4.25	.....
14 x 10, 14 x 9	3.75	4.25	$11.00
20 x 10, 18 x 10	4.00	4.50	11.00
16 x 10	4.00	4.50	11.00
14 x 8, 14 x 7	4.00	4.50	11.00
12 x 10	3.25	3.50	.....
12 x 8	3.25	3.50	9.25
12 x 7, 12 x 6	3.25	3.25	9.25
18 x 9, 16 x 9	4.00	4.50	11.00
16 x 8	4.00	4.50	11.50

To these prices add 20 per cent.

PEACH BOTTOM SLATE

Sizes	Price per Square F. O. B. Quarries
20 x 10, 18 x 10, 18 x 9	$5.60
16 x 9, 16 x 8	5.60
16 x 10, 16 x 11, 18 x 11	5.50
18 x 12, 20 x 11, 20 x 12	5.50
20 x 13, 22 x 11, 22 x 12	5.50
22 x 13, 22 x 14, 24 x 12	5.50
24 x 13, 24 x 14	5.35
24 x 15, 24 x 16	5.15
14 x 7, 14 x 8, 14 x 9, 14 x 10	5.25
12 x 6, 12 x 7, 12 x 8, 12 x 9, 12 x 10	4.75
11 x 5, 2-inch lap	3.50
11 x 6, 11 x 7, 11 x 8, 2-inch lap	3.75
10 x 5, 2-inch lap	3.25
10 x 6, 10 x 7, 10 x 8, 2-inch lap	3.50
Strictly 3-16 inch in thickness	7.00
Four to the inch in rick	7.50
Strictly ¼ inch in thickness	9.00

Drilling and countersinking, 50 cts. per square extra.

NO. 2

All sizes above 16 inch	$3.50
16 inch	3.40
14 inch	3.25
12 inch	2.75

The peach bottom slate is one of the best in the country; it is almost everlasting, never loses its color and is non-absorbent.

Add from 10 to 12 per cent to above prices.

SEA GREEN ROOFING SLATE

This is extensively used in many of the Western States. The color is not permanent, but it is strong and durable. For low-cost buildings it is a favorite in many localities and while the color changes, the durability of the material does not seem to suffer. It makes a good all-round slate roof.

Sizes	Price per Square F. O. B. Quarries	Sizes	Price per Square F. O. B. Quarries
24 x 14	$3.10	16 x 12	$3.00
24 x 12	3.10	16 x 10	3.00
22 x 14	3.00	16 x 9	3.00
22 x 12	3.10	16 x 8	2.90
22 x 11	3.20	14 x 10	2.90
20 x 12	3.10	14 x 9	2.90
20 x 10	3.20	14 x 8	2.90
18 x 12	3.10	14 x 7	2.70
18 x 10	3.10	12 x 8	2.70
18 x 9	3.10		

Add from 5 to 10 per cent to these prices.

To obtain the correct measurement of a surface of a slate when laid, and the number of squares on any particular surface, we simply subtract the lap from the length of the slate and half of the remainder will give the length of the surface exposed, which, when multiplied by the width of slate, gives the surface sought; so that to obtain the exact number of slates of any description required to cover any given surface is quite a simple matter. Further on I will give a rule for finding the number of slates required for covering any given area.

The following table gives the weight of slates of different thicknesses per square foot super.

Slate $\frac{3}{16}$ of an inch thick, 2.71 pounds per square foot.
Slate $\frac{1}{4}$ of an inch thick, 3.62 pounds per square foot.
Slate $\frac{3}{8}$ of an inch thick, 5.43 pounds per square foot.
Slate $\frac{1}{2}$ of an inch thick, 7.25 pounds per square foot.
Slate $\frac{5}{8}$ of an inch thick, 9.06 pounds per square foot.

Slate ¾ of an inch thick, 10.87 pounds per square foot.
Slate 1 inch thick, 14.5 pounds per square foot.
Slate 1¼ inches thick, 18.64 pounds per square foot.
Slate 1½ inches thick, 22.48 pounds per square foot.
Slate 2 inches thick, 30.00 pounds per square foot.

There are certain rules that are generally recognized by estimators and builders for the measurement of roofs, whether of slate, shingles or other materials, and may be given as follows:

For plain roofs, measure the length of the roof and multiply by the length of the rafter.

For roofs with hips, valleys, gables, dormers, etc., measure each section through the center and multiply by length of rafter, and in addition to the actual surface of the roof, measure the length of all hips and valleys, by one foot wide. No deduction is made for dormer windows, skylights, chimneys, etc., unless they measure more than 4 feet square. If more than 4 feet square, and less than 8 feet square, deduct one-half; if more than eight feet square, deduct the whole. If hips are mitered, charge extra. Ridge rolls, flashings, valleys, etc., are charged extra.

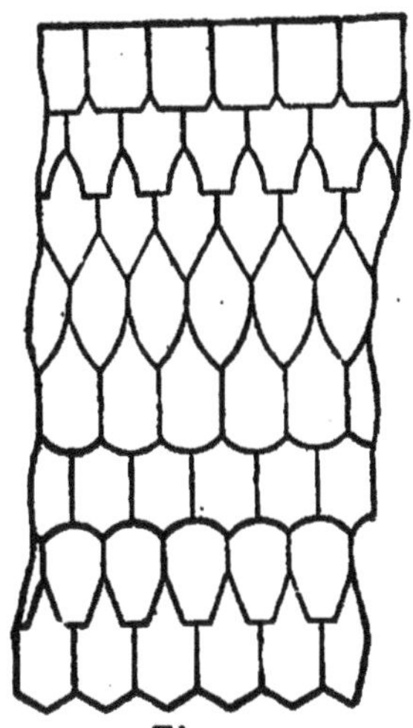

Fig. 17

The names given to ornamental slates and shingles are known by the trade and workmen as given and shown in Fig. 17; the examples are among those most used.

The expert slater, at the present writing, receives

$4.50 per day of nine hours, and he is supposed to lay about 1¼ squares, providing everything is handy for him. This wage, however, may be, and is, only given in a few localities. In some towns it is less, and in some cities it is more; so that the estimator should, whenever possible, obtain local prices both for labor and material; then he cannot well make many errors in his estimate if he is at all careful.

The following hints regarding preparing estimates may not be out of place, and I give them even if I risk being accused of repetition, as I have given nearly the same advice in previous pages; but it cannot be given too often to the young and inexperienced estimator, for the omission of a single item may result in grievous loss.

The first thing to do, before commencing to make an estimate of the cost of a job of work, is to see to it that the drawings and specifications be carefully studied and remarks made for future guidance or reference.

Excavator and Bricklayer.—Take the dimensions for the excavations wholly from plans and sections, then refer to the specification and add whatever does not appear on the drawings.

Take brick work as above directed.

The chimney bars may be taken with the dimensions of the breasts; refer to specifications for description.

The centering and spring pieces may be taken with arches.

The wood bricks and springing pieces may be taken with brick arches.

Take dimensions by the foot run of the making good and restorings of all stone sills, after mason, etc.

Slater.—Take from the plan of roof and section, then refer to specification and clear all off.

In the bill provide for leaving all slating perfect at completion of the work.

Carpenter.—Take from plans and sections, referring to specification. Take all iron attached to timbers. Find the quantity of boarding, or battening, to roof, with the slating, deducting eaves, cuttings, etc., from the latter; if much discrepancy, there must be an error.

Plumber.—Take everything from specification, referring to drawings only for lengths. Be careful in allowing all turnings up under slating and against walls, round rolls, flashing, etc. (refer to the quantity of boarding to flats, etc., as a check.)

Mason.—Take from specification, referring to the drawings only for dimensions. Attend to the cube quantities, scantling lengths, etc., also to the method pointed out for taking labor.

Joiner and Hardware.—Peruse specification, referring to drawings only for dimensions. Take hardware from floor to floor. Provide for casing stairs, and covering handrail to prevent any injury during the progress of the work, and for projecting masonry in like manner.

Provide for attending plumbers to sinks, cisterns, W. C., etc., stating how many of each.

Plasterer, Internally.—Look carefully to specification, particularly as to enrichments, referring to drawing for dimensions. Whitening and coloring is taken from plastering, but appears separately.

Provide for making good round mantels after mason.

Glazier.—Can find all in specification, referring to drawings for any size. Check quantity of glass by the sashes, allowing for wood. Provide for leaving same clean and sound.

Painter.—All taken from specification.

All wood work painted may be collected from the joiner; one-seventh for edges; when both sides are painted, double dimension. Painting for plastered walls, from plasterer.

Paper-Hanger.—May be taken from plasterer.

Summary of trades in order. Conditions of contract to be taken from specification, and furnished in the memorandum sheet.

Fees.—Government, municipal, sewer, and architect's fees to be attached at end.

At the head of each trade give fair description from specification of quality of materials, etc.

Lastly.—Generally review the whole of the drawings and specifications, that nothing may be omitted or misrepresented.

RULES, TABLES, NOTES, DATA, AND POINTERS USEFUL TO THE ESTIMATOR

The following tables, data, etc., have been specially selected for the use of the estimator, and will be found useful for reference and for making hurried approximate estimates of work in detail or in bulk. The items are carefully indexed, so that any particular one of them may be found without much loss of time; a matter of considerable importance to the busy man. It would extend the limits of this book too far beyond the size intended to insert tables of scantling measurement, wages, extended tables of diameters, circumferences and areas of circles and similar matter, which after all are not of much actual service to the estimator, but which are usually published in works of this kind.

The average weight of medium and heavy cast-iron drain pipes are given in the following tables, viz.:—

WEIGHT AND THICKNESS OF CAST-IRON DRAIN-PIPES

Diameter of Pipe	Length exclusive of socket	Thickness of Metal	Depth of Socket	Average weight per pipe			Approximate weight per foot run		
in.	ft.	in.	in.	cwt.	qr.	lb.	cwt.	qr.	lb.
4	9	11/32	4	1	1	12	0	0	17
	9	3/8	4	1	1	20	0	0	18
4½	9	3/8	4	1	2	14	0	0	20
5	9	3/8	4	1	3	0	.0	0	22
	9	13/32	4	2	0	0	0	0	25
6	9	13/32	4¼	2	1	0	0	1	0
	9	7/16	4¼	2	2	0	0	1	3
	9	½	4¼	2	3	7	0	1	7
7	9	7/16	4¼	2	3	14	0	1	8
	9	½	4¼	3	1	0	0	1	12
8	9	7/16	4¼	3	1	0	0	1	12
	9	½	4¼	3	2	24	0	1	18
9	9	½	4½	4	0	7	0	1	23
	9	17/32	4½	4	2	0	0	2	0
10	9	½	4½	4	2	0	0	2	0
	9	17/32	4½	5	0	0	0	2	6
12	9	17/32	4½	5	2	24	0	2	15
	9	9/16	4½	6	0	0	0	2	19
	9	5/8	4½	6	2	14	0	2	26
	12	17/32	4½	7	2	7	0	2	14
	12	9/16	4½	7	3	14	0	2	18
	12	5/8	4½	8	2	14	0	2	25

TABLE SHOWING NUMBER OF BRICKS IN WALLS OF VARIOUS THICKNESSES

Per sq. foot of wall	THICKNESS OF WALLS									
	4½ in or ½ brick	9 in. or 1 brick	13 in. or 1½ brick	18 in. or 2 brick	22 in. or 2½ brick	26 in. or 3 brick	30 in. or 3½ brick	35 in. or 4 brick	39 in. or 4½ brick	44 inch or 5 brick
ft. in.										
0.6	3½	7	10½	14	17½	21	24½	28	31½	35
1.0	7	14	21	28	35	42	49	56	63	70
1.6	10½	21	31½	42	52½	63	73½	84	94½	105
2.0	14	28	42	56	70	84	98	112	126	140
2.6	17½	35	52½	70	87½	105	122½	140	157½	175
3.0	21	42	63	84	105	126	147	168	189	210
3.6	24½	49	73½	98	122½	147	171½	196	220½	245
4.0	28	56	84	112	140	168	196	224	252	280
4.6	31½	63	94½	126	157½	189	220½	252	283½	315
5.0	35	70	105	140	175	210	245	280	315	350
5.6	38½	77	115½	154	192½	231	269½	308	346½	385
6.0	42	84	126	168	210	252	294	336	378	420
6.6	45½	91	136½	182	227½	273	318½	364	409½	455
7.0	49	98	147	196	245	294	343	392	441	490
7.6	52½	105	157½	210	262½	315	367½	420	472½	525
8.0	56	112	168	224	280	336	392	448	504	560
8.6	59½	119	178½	238	297½	357	416½	476	535½	595
9.0	63	126	189	252	315	378	441	504	567	630
9.6	66½	133	199½	266	332½	399	465½	532	598½	665
10.0	70	140	210	280	350	420	490	560	630	700
15.0	105	210	315	420	525	630	735	840	945	1050
20.0	140	280	420	560	700	840	980	1120	1260	1400
30.0	210	420	630	840	1050	1260	1470	1680	1890	2100
40.0	280	560	840	1120	1400	1680	1960	2240	2520	2800
50.0	350	700	1050	1400	1750	2100	2450	2800	3150	3500
60.0	420	840	1260	1680	2100	2520	2940	3360	3780	4200
70.0	490	980	1470	1960	2450	2940	3430	3920	4410	4900
80.0	560	1120	1680	2240	2800	3360	3920	4480	5040	5600
90.0	630	1260	1890	2520	3150	3780	4410	5040	5670	6300
100.0	700	1400	2100	2800	3500	4200	4900	5600	6300	7000
200.0	1400	2800	4200	5600	7000	8400	9800	11200	12600	14000
300.0	2100	4200	6300	8400	10500	12600	14700	16800	18900	21000
400.0	2800	5600	8400	11200	14000	16800	19600	22400	25200	28000
500.0	3500	7000	10500	14000	17500	21000	24500	28000	31500	35000
600.0	4200	8400	12600	16800	21000	25200	29400	33600	37800	42000
700.0	4900	9800	14700	19600	24500	29400	34300	39200	44100	49000
800.0	5600	11200	16800	22400	28000	33600	39200	44800	50400	56000
900.0	6300	12600	18900	25200	31500	37800	44100	50400	56700	63000
1000.0	7000	14000	21000	28000	35000	42000	49000	56000	63000	70000

Brick work is generally measured by 1,000 bricks laid in the wall. In consequence of variations in size of bricks, no rule for volume of laid brick can be exact. The following scale is, however, a fair average:

7 compressed bricks to a super. foot 4-in. wall.
14 compressed bricks to a super. foot 9-in. wall.
21 compressed bricks to a super. foot 13-in. wall.
28 compressed bricks to a super. foot 18-in. wall.
35 compressed bricks to a super. foot 22-in. wall.

Corners are not measured twice, as in stone work. Openings over 2 feet square are deducted. Arches are counted from the spring. Fancy work counted 1½ bricks for 1. Pillars are measured on their face only.

A cubic yard of mortar requires 1 cubic yard of sand and 9 bushels of lime, and will fill 30 hods.

One thousand bricks closely stacked occupy about 56 cubic feet.

One thousand old bricks, cleaned and loosely stacked, occupy about 72 cubic feet.

One superficial foot of gauged arches requires 10 bricks.

Pavements, according to size of bricks, take 38 brick on flat and 60 brick on edge per square yard. on an average.

Five courses of brick will lay 1 foot in height on a chimney; 6 bricks in a course will make a flue 4 inches wide and 12 inches long, and 8 bricks in a course will make a flue 8 inches wide and 16 inches long.

SAFE BEARING LOADS

BRICK AND STONE MASONRY

Brickwork—	Lbs. per sq. in.
Bricks, hard, laid in lime mortar	100
Hard, laid in Portland cement mortar	200
Hard, laid in Rosendale cement mortar	150
Masonry—	
Granite, capstone	700
Squared stonework	350
Sandstone, capstone	350
Squared stonework	175
Rubble stonework, laid in lime mortar	80
Rubble stonework, laid in cement mortar	150
Limestone, capstone	500
Squared stonework	250
Rubble, laid in lime mortar	80
Rubble, laid in cement mortar	150
Concrete, 1 Portland, 2 sand, 5 broken stone	150
Foundation Soils—	Tons per sq. foot.
Rock, hardest in native bed	100—
Equal to best Ashlar masonry	25–40

Foundation Soils—	Tons per sq. foot.
Equal to best brick	15–20
Clay, dry, in thick beds	4–6
Moderately dry, in thick beds	2–4
Soft	1–2
Gravel and coarse sand, well cemented	8–10
Sand, compact and well cemented	4–6
Clean, dry	2–4
Quicksand, alluvial soils, etc	.5–1

EXCAVATIONS

Excavations are measured by the yard (27 cubic feet), and irregular depths or surfaces are generally averaged in practice.

MASONRY

Stone masonry is measured by two systems, Quarryman's and Mason's Measurements.

By the Quarryman's Measurements the actual contents are measured; that is, all openings are taken out and all corners are measured single.

By Mason's Measurements, corners and piers are doubled, and no allowance made for openings less than 3 × 5 feet and only half the amount of openings larger than 3 × 5 feet.

Range work and cut work is measured superficially and in addition to wall measurement.

An average of six bushels of sand and cement per perch of Rubble Masonry.

Stone walls are measured by the perch (24¾ cubic feet). Openings less than 3 feet wide are counted solid; over 3 feet deducted, but 18 inches are added to the running measure for each jamb built.

Arches are counted solid from their spring. Corners of buildings are measured twice. Pillars less than 3 feet are counted on 3 sides as lineal, multiplied by fourth side and depth.

It is customary to measure all foundation and dimension stone by the cubic foot. Water tables and base courses by lineal feet. All sills and lintels or ashlar by superficial feet, and no wall less than 18 inches thick.

The height of brick or stone piers should not exceed 12 times their thickness at the base.

Masonry is usually measured by the perch (containing 24.75 cubic feet), but in practice 25 cubic feet are considered a perch of masonry.

Concreting is usually measured by the cubic yard (27 cubic feet).

A cord of stone, 3 bushels of lime, and a cubic yard of sand, will lay 100 cubic feet of wall.

Cement, 1 bushel, and sand, 2 bushels, will cover 3½ square yards 1 inch thick; 4½ square yards ¾ inch thick, and 6¾ square yards ½ inch thick; 1 bushel of cement and 1 of sand will cover 2¼ square yards 1 inch thick, 3 square yards ¾ inch thick, and 4½ square yards 2 inch thick.

THE PROPORTION OF STOCK BRICKS AND MORTAR TO A ROD OF BRICKWORK

Thickness of Mortar Joints	Gauge or Height of 4 Courses	Cubic Feet of Bricks	Cubic Feet of Mortar	Number of Bricks
inch.				
¼	12½	258	58	4180
	12	257	59	4350
	11½	256	60	4540
⅜	12½	237	79	4010
	12	236	80	4176
	11½	234	82	4358

Bricks absorb about $\frac{1}{15}$ of their weight of water.

A bricklayer's hod measures 16 in. x 9 in. x 9 in. = 1296 cubic inches.

Ditto will hold 20 bricks.

Ditto, ditto ⅔ cubic foot of mortar.

Ditto, ditto ½ bushel nearly.

The proportions of lime, sand, or cement required for a rod of brickwork are:

Of white stone lime	26	Cubic Feet.
Sand	78	
Gray lime	36	Cubic Feet.
Sand	72	
Blue lime	38	Cubic Feet.
Sand	77	
Roman or Portland cement	45	Cubic Feet.
Sand	45	

One rod of brickwork requires 126 gallons of water to slake the lime and mix the mortar.

A load of Mortar = 1 cubic yard, and will fill 30 hods.

	Mortar produced in cubic feet.
1 imperial bushel of blue lime, unslaked, weighing 70 lbs; 2 imperial bushels of sand, weighing 103 lbs; 6½ gallons of water	2.75
1 imperial bushel of blue lime, unslaked; 3 imperial bushels of sand; 7½ gallons of water	3.25
1 imperial bushel of Portland cement, weighing 99 lbs; 1 imperial bushel of sand, weighing 103 lbs; 3¾ gallons of water	1.75
1 imperial bushel of Portland cement; 2 imperial bushels of sand; 5¼ gallons of water	2.58
1 imperial bushel of Portland cement; 3 imperial bushels of sand; 6¾ gallons of water	3.42
1 imperial bushel of Roman cement, weighing 72 lbs; 9½ gallons of water	1.125

Note:—The mortar produced weighed 106 lbs.

1 imperial bushel of Roman cement; 1 imperial bushel of sand (103 lbs); 9½ gallons of water	1.764

Note:—The mortar weighed 196 lbs.

	Concrete produced in cubic feet
1 imperial bushel of Portland cement	2.08
1 imperial bushel of stone, broken small	
½ imperial bushel of sand	
4½ gallons of water	

Lime and sand, and cement and sand lose about one-third of their bulk when made into mortar.

Lime, or Portland cement, and sand require to mix into mortar about one-third of their bulk of water.

Brick nogging requires—

Per yard superficial, 45 stock bricks laid flat.

Per yard superficial, 30 stock bricks on edge.

Per yard superficial, ¾ cubic foot mortar when flat.

Per yard superficial, ½ cubic foot mortar on edge.

THE NUMBER OF BRICKS AND QUANTITY OF BRICKWORK IN WELLS AND CYLINDRICAL SEWERS FOR EACH FOOT IN DEPTH OR LENGTH

	Half Brick Thick			One Brick Thick		
	Number of Bricks		Cubic Feet of Brick-work	Number of Bricks		Cubic Feet of Brick-work
	Laid Dry	Laid in Mortar		Laid Dry	Laid in Mortar	
1.0	28	23	1.6198	70	58	4.1233
1.3	33	27	1.8145	80	66	4.7124
1.6	38	31	2.2089	90	74	5.3015
1.9	43	35	2.5035	102	82	5.8905
2.0	48	41	2.7979	112	92	6.4795
2.3	53	44	3.0926	122	100	7.0686
2.6	58	48	3.3870	132	108	7.6577
3.0	68	57	3.9760	154	126	8.8357
3.6	79	65	4.5651	174	142	10.0139
4.0	89	73	5.1541	194	159	11.1919
4.6	100	82	5.7432	214	176	12.3701
5.0	110	90	6.3322	234	192	13.5481
5.6	120	98	6.9213	254	209	14.7263
6.0	130	107	7.5103	276	226	15.9043
6.6	140	115	8.0994	296	242	17.0825
7.0	150	123	8.6884	316	260	18.2605
7.6	160	131	9.2775	336	276	19.4387
8.0	170	140	9.8665	358	292	20.6167
8.6	180	148	10.4556	378	308	21.7949
9.0	191	156	11.0446	398	326	22.9729
10.0	212	174	12.2227	438	360	25.3291

THE THICKNESS OF WALLS FOR DWELLING HOUSES—BRICK

Maximum Height = 100 feet.
Maximum Length.

45 feet.	80 feet.	Unlimited.
Inches	Inches	Inches
Two stories of 21½	Two stories of 26	One story of 30
Three stories of 17½	Two stories of 21½	Two stories of 26
Remainder.... 13	Two stories of 17½	Two stories of 21½
	Remainder.... 13	Two stories of 17½
		Remainder.... 13

Maximum Height = 90 feet.
Maximum Length.

45 feet.	70 feet.	Unlimited.
Inches	Inches	Inches
Two stories of 21½	One story of 26	One story of 30
Two stories of 17½	Two stories of 21½	Two stories of 26
Remainder.... 13	Two stories of 17½	One story of 21½
	Remainder.... 13	Two stories of 17½
		Remainder.... 13

Maximum Height = 80 feet.
Maximum Length.

40 feet.	60 feet.	Unlimited.
Inches	Inches	Inches
One story of 21½	Two stories of 21½	One story of 26
Two stories of 17½	Two stories of 17½	Two stories of 21½
Remainder.... 13	Remainder.... 13	Two stories of 17½
		Remainder.... 13

Maximum Height = 70 feet.
Maximum Length.

40 feet.	55 feet.	Unlimited.
Inches	Inches	Inches
Two stories of 17½	One story of 21½	One story of 26
Remainder.... 13	Two stories of 17½	Two stories of 21½
	Remainder.... 13	One story of 17½
		Remainder.... 13

Maximum Height = 60 feet.
Maximum Length.

30 feet.	50 feet.	Unlimited.
Inches	Inches	Inches
One story of 17½	Two stories of 17½	One story of 21½
Remainder.... 13	Remainder.... 13	Two stories of 17½
		Remainder.... 13

Maximum Height = 50 feet.
Maximum Length.

30 feet	45 feet.	Unlimited.
Inches	Inches	Inches
Wall below the topmost story 13	One story of 17½	One story of 21½
Topmost story 8½	Rest of wall below topmost story 13	One story of 17½
Remainder.... 8½	Topmost story 8½	Remainder.... 13
	Remainder.... 8½	

Maximum Height = 40 feet.
Maximum Length.

35 feet.	Unlimited.
Inches	Inches
Wall below two topmost stories 13	One story of.............17½
Two topmost stories of.... 8½	Rest of wall below topmost story.................13
Remainder............... 8½	Topmost story 8½
	Remainder.............. 8½

Maximum Height = 30 feet.
Maximum Length.

35 feet.	Unlimited.
Inches	Inches
Wall below two topmost stories................. 13	Wall below topmost story..13
Two topmost stories....... 8½	Topmost story 8½
Remainder................ 8½	Remainder............... 8½

Maximum Height = 25 feet.
Maximum Length.

30 feet.	Unlimited.
Inches	Inches
From base to top of wall... 8½	Wall below topmost story..13
	Topmost story............ 8½
	Remainder.............. 8½

THE THICKNESS OF WALLS FOR WAREHOUSES—BRICK

Maximum Height in feet	Maximum Length in feet	Thickness at Base in inches	Maximum Length in feet	Thickness at Base in inches	Maximum Length in feet	Thickness at Base in inches
100	55	26	70	30	Length unlimited	34
90	60	26	70	30		34
80	45	21½	60	26		30
70	30	17½	45	21½		26
60	35	17½	50	21½		26
50	40	17½	70	21½		26
40	30	13	60	17½		21½
30	45	13	—	—		17½
25	—	—	—	—		13

The thickness of the walls at the top for warehouses, and for 16 feet below the top, shall = 13 inches; and the intermediate parts of the wall, between the base and such 16 feet below the top, to be solid throughout the space between straight lines drawn on each side of the wall from the base to the part 16 feet below the top, as above determined; but in walls not exceeding 30 feet in height, those of the topmost story may be 8½ inches thick.

The thickness to be increased to one-sixteenth part of the height of the story for dwelling houses, and to one-fourteenth part for warehouses, in case the thickness determined by the foregoing tables be less than that proportion.

The width of the footings at the base to be *double* the thickness of the wall, to diminish in regular offsets, and to be equal in height to one-half of the width at base.

ROOFS GENERALLY

SHINGLING

To find the number of shingles required to cover 100 square feet deduct 3 inches from the length, divide the remainder by 3, the result will be the exposed length of a shingle; multiplying this by the average width of a shingle, the product will be the exposed area. Dividing 14,400, the number of square inches in a square, by the exposed area of a shingle will give the number required to cover 100 square feet of roof.

In estimating the number of shingles required, an allowance should always be made for waste.

Estimates on cost of shingle roofs are usually given per 1,000 shingles.

TABLE FOR ESTIMATING SHINGLES

Length of Shingles	Exposure to Weather Inches	No. of sq. ft. of Roof Covered by 1000 Shingles.		No. of Shingles Required for 100 sq. feet of Roof.	
		4 in. Wide	6 in. Wide	4 in. Wide	6 in. Wide
15 in.	4	111	167	900	600
18	5	139	208	720	480
21	6	167	250	600	400
24	7	194	291	514	343
27	8	222	333	450	300

SLATING

A square of slate or slating is 100 superficial feet.

In measuring, the width of eaves is allowed at the widest part. Hips, valleys and cuttings are to be measured lineal, and 6 inches extra is allowed.

The thickness of slates required is from 3-16 to 5-16 of an inch, and their weight varies when lapped from 4.5 to 6¾ pounds per square foot.

The "laps" of slates vary from 2 to 4 inches, the standard assumed to be 3 inches.

TO COMPUTE THE NUMBER OF SLATES OF A GIVEN SIZE REQUIRED PER SQUARE

Subtract 3 inches from the length of the slate, multiply the remainder by the width and divide by 2. Divide 14,400 by the number so found and the result will be the number of slates required.

TABLE SHOWING NUMBER OF SLATES AND POUNDS OF NAILS REQUIRED TO COVER 100 SQUARE FEET OF ROOF

Size of Slate	Length of Exposure	No. Required	Nails Required
14 in. x 28 in.	12½ in.	83	.6 lbs.
12 x 24	10½	114	.833
11 x 22	9½	138	1.
10 x 20	8½	165	1.33
9 x 18	7½	214	1.5
8 x 16	6½	277	2.
7 x 14	5½	377	2.66
6 x 12	4½	533	3.8

APPROXIMATE WEIGHT OF MATERIALS FOR ROOFS

Material	Average weight lb. per sq. ft.
Corrugated galvanized iron No. 20, unboarded	2¼
Copper, 16 oz. standing seam	1¼
Felt and asphalt, without sheathing	2
Glass, ⅛ inch thick	1¾
Hemlock sheathing, 1 inch thick	2
Lead, about ⅛ inch thick	6 to 8
Lath-and-plaster ceiling (ordinary)	6 to 8
Mackite, 1 inch thick, with plaster	10
Neponset roofing felt, 2 layers	½
Spruce sheathing, 1 inch thick	2½
Slate, 3/16 inch thick, 3-inch double lap	6¾
Slate, ⅛ inch thick, 3-inch double lap	4½
Shingles, 6″ × 18″, ⅓ to weather	2
Skylight of glass, 3/16 to ½ inch, including frame	4 to 10
Slag roof, 4-ply	4

Material	Average weight lb. per sq. ft.
Terne plate, IC, without sheathing	$\frac{1}{2}$
Terne Plate, IX, without sheathing	$\frac{5}{8}$
Tiles (plain), $10\frac{1}{2}'' \times 6\frac{1}{4}'' \times \frac{5}{8}''$ — $5\frac{1}{4}''$ to weather	18
Tiles (Spanish), $14\frac{1}{2}'' \times 10\frac{1}{2}''$ — $7\frac{1}{4}''$ to weather	$8\frac{1}{2}$
White-pine sheathing, 1 inch thick	$2\frac{1}{2}$
Yellow-pine sheathing, 1 inch thick	4

SNOW AND WIND LOADS

Data in regard to snow and wind loads are necessary in connection with the design of roof trusses.

Snow Load.—When the slope of a roof is over 12 inches rise per foot of horizontal run, a snow and accidental load of 8 pounds per square foot is ample. When the slope is under 12 inches rise per foot of run, a snow and accidental load of 12 pounds per square foot should be used. The snow load acts vertically, and therefore should be added to the dead load in designing roof trusses. The snow load may be neglected when a high wind pressure has been considered, as a great wind storm would very likely remove all the snow from the roof.

Wind Load.—The wind is considered as blowing in a horizontal direction, but the resulting pressure upon the roof is always taken *normal* (at right angles) to the slope. The wind pressure against a vertical plane depends on the velocity of the wind, and, as ascertained by the United States Signal Service at Mount Washington, N. H., is as follows:

Velocity (Mi. per Hr.)	Pressure (Lb. per Sq. Ft.)	
10	0.4	Fresh breeze.
20	1.6	Stiff breeze.
30	3.6	Strong wind.
40	6.4	High wind.
50	10.0	Storm.
60	14.4	Violent storm.
80	25.6	Hurricane.
100	40.0	Violent hurricane.

The wind pressure upon a cylindrical surface is one-half that upon a flat surface of the same height and width.

Since the wind is considered as traveling in a horizontal direction, it is evident that the more nearly vertical the slope of the roof, the greater will be the pressure, and the more nearly horizontal the slope, the less will be the pressure. The following table gives the pressure exerted upon roofs of different slopes, by a wind pressure of 40 pounds per square foot on a vertical plane, which is equivalent in intensity to a violent hurricane.

WIND PRESSURES ON ROOFS

(Pounds per Square Foot)

Rise In. per Foot of Run	Angle with Horizontal	Pitch Proportion of Rise to Span	Wind Pressure Normal to Slope
4	18° 25′	1/6	16.8
6	26° 33′	1/4	23.7
8	33° 41′	1/3	29.1
12	45″ 0′	1/2	36.1
16	53° 7′	2/3	38.7
18	56° 20′	3/4	39.3
24	63° 27′	1	40.0

In addition to wind and snow loads upon roofs, the weight of the principals or roof trusses, including the other features of the construction, should be figured in the estimate. For light roofs having a span of not over 50 feet, and not required to support any ceiling, the weight of the steel construction may be taken at 5 pounds per square foot; for greater spans, 1 pound per square foot should be added for each 10 feet increase in the span.

COMPARATIVE COST OF ROOFS

It often happens that an estimator is asked as to the difference in the cost of roofs, and on his answer the construction of the work may depend; therefore it is necessary that he should be able to give his answer with some degree of intelligence and exactness; and the following, to some extent, will enable him to do this.

For instance, take a "span roof," by which we mean one having two sides inclining to a ridge, and let the length of the rafter be 16 feet, and that of the roof from edge to edge be 14 feet.

Then it contains on each side a trifle over 7 squares of 100 superficial feet each.

If the roof is to be slated or tinned it will require the sheathing to be laid close, and with what is called "match mill-planed timber," which is provided with tongue and groove, and need not, as the name implies, be mill-planed, although it usually is.

We next come to consider the cost of sheathing, nails, and labor required in putting it on, which, approximately, is as follows:

PREPARING FOR SLATE OR TIN ROOF

7 squares of roofing require 700 feet of sheathing at $15 per M	$10.50
Labor required in putting same on, at 50 cents per square	3.50
Nails for fastening sheathing boards	1.00
Total cost	$15.00

SLATE ROOF

We find the cost of the slate roof to wit:

For preparing for roof	$15.00
For 7 squares of slating, including labor, material, etc., at $12 per square	84.00
Total cost	$99.00

Thus it will be seen that the total cost of 7 squares of slating aggregates a cost of $99.00, or $14.15 per square.

TIN ROOF

Since the work of preparing for the tin roof is the same as for slate, we add to it the cost for tin and painting as follows:

For preparing for roof	$15.00
For 7 squares of tin work at 75 cents per square, including material and labor	52.50
For 78 yards of paint, 2-coat work, at 15 cents per yard	11.70
Total cost	$79.20

At these figures we find that 7 squares of tin roofing will cost $79.20, or a trifle over $11.31 per square.

SHINGLE ROOF

In estimating the amount of sheathing required for a shingle roof, we bear in mind the fact that it will not be necessary to lay boards close together; but strips 3 inches wide can be used, and if so, it will require about one-half of the amount it does when laid close, as for the slate or tin roof. Hence the following is the approximate estimate of cost.

300 feet of sheathing at $12.50 per M	$ 3.75
Labor required in putting same on	1.05
Nails for sheathing, etc	.50
7,000 shingles, nails and labor at $7 per square	49.00
Total cost	$54.30

Thus the cost of 7 squares of shingling will aggregate $54.30, or a trifle over $7.75 per square.

COMPOSITION ROOF

Now suppose that the slope of the roof permitted the surface to be covered with gravel or composition roofing, then the sheathing need not be laid as carefully as for tin or slate, and an inferior quality of lumber can be used; the only requirements being that the surface must be level and smooth.

In such a case the estimate of cost would be as follows:

700 feet of sheathing at $12.50 per M	$ 8.75
Putting on same at 35 cents per square	2.45
Nails for sheathing, etc	1.00
7 squares roofing material, etc., $4 per square	28.00
Total cost	$40.20

Making the cost of 7 squares amount to $40.20, or a trifle over $5.74 per square.

Slate on iron purlins	$2.00 to $7.00	per sq.
Metal tile, tin	8.50 to 9.75	per sq.
Metal tile, steel, lead-coated	10.75 to 13.75	per sq.
Rubber roofing	2.00 to 3.75	per sq.
Felt and gravel	6.50	per sq.
Ornamental tile	40.00 to 60.00	per M.
Tile shingles	21.00 to 35.00	per M.
Charcoal tin plates, I.C., 14×20 ins.	6.00 to 6.50	per box of 112.
Charcoal tin plates, I.C., 20×28 ins.	12.00 to 13.00	per box of 112.
Charcoal tin plates, I.X., 14×20 ins.	7.50 to 8.50	per box of 112.
Charcoal tin plates, I.X., 20×28 ins.	15.00 to 17.00	per box of 112.
Coke plates, tin, I.C., 14×20 ins.	5.50	per box of 112.
Coke plates, tin, I.C., 20×28 ins	11.50 to 12.00	per box of 112.
Coke plates, tin, I.X., 14×20 ins.	7.50	per box of 112.
Charcoal plate, terne, I.C., 14×20 ins	5.50	per box of 112.
Charcoal plate, terne, I.C., 20×28 ins	10.75 to 11.00	per box of 112.
Charcoal plate, terne, I.X., 14×20 ins.	6.40	per box of 112.
Charcoal plate, terne, I.X., 20×28 ins.	12.80	per box of 112.

FLAT SEAM TIN ROOFING

Table showing quantity of 14″ x 20″ tin required to cover a given number of square feet with flat seam tin roofing. A sheet of 14″ x 20″ with ½″ edges measures, when edged or folded, 13″ x 19″ or 247 square inches. In the following, all fractional parts of a sheet are counted a full sheet.

No. of sq. feet	Sheets required	No. of sq. feet	Sheets required	No. of sq. feet	Sheets required	No. of sq. feet	Sheets required	No. of sq. feet	Sheets required
100	59	280	164	460	269	640	374	820	479
110	65	290	170	470	275	650	379	830	484
120	70	300	175	480	280	660	385	840	490
130	76	310	181	490	286	670	391	850	496
140	82	320	187	500	292	680	397	860	502
150	88	330	193	510	298	690	403	870	508
160	94	340	199	520	304	700	409	880	514
170	100	350	205	530	309	710	414	890	519
180	105	360	210	540	315	720	420	900	525
190	111	370	216	550	321	730	426	910	531
200	117	380	222	560	327	740	432	920	537
210	123	390	228	570	333	750	438	930	543
220	129	400	234	580	339	760	444	940	549
230	135	410	240	590	344	770	449	950	554
240	140	420	245	600	350	780	455	960	560
250	146	430	251	610	356	790	461	970	566
260	152	440	257	620	362	800	467	980	572
270	158	450	263	630	368	810	473	990	578

1000 square feet 583 sheets.

A box of 112 sheets 14″×20″ will cover approximately 192 square feet.

STANDING SEAM TIN ROOFING

Table showing quantity of 20″×28″ tin required to cover a given number of square feet with standing seam roofing. The standing seams and the locks on a steep roof require 2¾″ off the width and ¾″ off the length of the sheet; fractional parts are counted as a full sheet. A sheet will cover 475 square inches.

No. of sq. feet	Sheets required	No. of sq. feet	Sheets required	No. of sq. feet	Sheets required	No. of sq. feet	Sheets required	No. of sq. feet	Sheets required
100	31	280	85	460	140	640	194	820	249
110	34	290	88	470	143	650	197	830	252
120	37	300	91	480	147	660	200	840	255
130	40	310	94	490	149	670	203	850	258
140	43	320	97	500	152	680	206	860	261
150	46	330	100	510	158	690	209	870	264
160	49	340	103	520	161	700	212	880	267
170	52	350	106	530	164	710	215	890	270
180	55	360	109	540	167	720	218	900	273
190	58	370	112	550	170	730	221	910	276
200	61	380	115	560	173	740	224	920	279
210	64	390	118	570	176	750	228	930	282
220	67	400	122	580	182	760	231	940	285
230	70	410	125	590	184	770	234	950	288
240	73	420	128	600	185	780	237	960	291
250	76	430	131	610	185	790	240	970	294
260	79	440	134	620	188	800	243	980	297
270	82	450	137	630	191	810	246	990	300

1000 square feet 303 sheets.

A full box 112 sheets 20″×28″ will cover approximately 370 square feet.

It must be understood that the figures given in the foregoing are not considered as being correct or suited to all localities; they may be taken as approximately exact, but in all cases the percentage of difference in cost may be taken as fairly correct, and it is this result for which the tables were prepared.

SPECIFIC GRAVITY AND WEIGHTS

BUILDING MATERIALS

Name of Material	Weight per Cu. ft. lb.	Specific Gravity
Brick, pressed	150	2.40
Brick, common	125	2.00
Cement, Portland	80 to 100	1.44
Cement, Rosedale	56	.89
Common brickwork, cement mortar	130	2.10
Common brickwork, lime mortar	120	1.90
Concrete cement	140	2.25
Earth, dry, shaken	82 to 92	1.36
Earth, rammed	90 to 100	1.52
Glass, window	157	2.52
Granite	170	2.72
Granite or limestone, rubble work	138	2.21
Granite or limestone, well dressed	165	2.65
Limestones and marbles	168	2.70
Lime, Quick	53	.85
Mortar, hardened	103	1.65
Plaster of paris	141.6	2.27
Pressed brickwork	140	2.25
Sand	90 to 106	2.65
Sandstone	151	2.41
Shales	162	2.60
Slate	175	2.80
Trap Rock	187	3.00

WOODS (DRY)

Name of Material	Weight Per ft. Bm.	Weight per Cu. ft. lb.	Specific Gravity
Ash	3.9	47	.752
Ash, American, white	3.2	38	.610
Boxwood	5.	60	.960
Cherry	3.5	42	.672
Chestnut	3.4	41	.660
Cork	1.3	15	.250

Elm	2.9	35	.560
Ebony	6.3	76.1	1.220
Hemlock	2.1	25	.400
Hickory	4.4	53	.850
Lignum Vitæ	6.9	83	1.330
Mahogany, Spanish	4.4	53	.850
Mahogany, Honduras	2.9	35	.560
Maple	4.1	49	.790
Oak, live	4.9	59.3	.950
Oak, white	4.0	48	.770
Oak, red	3.2	40	.640
Pine, white	2.1	25	.400
Pine, yellow	2.8	34.3	.550
Pine, southern	3.7	45	.720
Sycamore	3.1	37	.590
Spruce	2.1	25	.400
Walnut	3.2	38	.610

The estimated weight of logs is one-half more than the estimated weight of the green lumber of the same kind of wood.

THE METRIC SYSTEM

The metric system is based on the meter, which, according to the United States Coast and Geodetic Survey Report of 1884, is equal to 39.370432 inches. The value commonly used is 39.37 inches, and is authorized by the United States government. The meter is defined as one ten-millionth the distance from the pole to the equator, measured on a meridian passing near Paris.

There are three principal units: the meter, the liter (pronounced lee-ter), and the gram, the units of length, capacity and weight, respectively. Multiples of these units are obtained by prefixing to the names of the principal units the Greek words Deca (10), hecto (100), and kilo (1,000); the submultiples, or divisions, are obtained by prefixing the Latin words Deci (1/10), centi (1/100), and milli (1/1000). These prefixes form the key to the entire system. In the following tables the abbreviations of the principal units of these sub-multiples begin with a small letter, while those of the multiples begin with a capital letter; they should always be written as here printed.

MEASURES OF LENGTH

Name	Meters	U. S. In.	Feet
Millimeter (mm.)	= .001	= .039370	= .003281
Centimeter (cm.)	= .010	= .393704	= .032809
Decimeter (dm.)	= .100	= 3.937043	= .328087
Meter (m.)	= 1.000	= 39.370432	= 3.380869
Decameter (Dm.)	= 10.000		= 32.808690
Hectometer (Hm.)	= 100.000		= 328.086900
Kilometer (Km.)	= 1,000.000	= .621 mi.	= 3,280.869000
Myriameter (Mm.)	= 10,000.000	= 6.214 mi.	= 32,808.690000

The centimeter, meter and kilometer are the units in practical use, and may be said to occupy the same position in the metric system as do inches, yards and miles in the United States and English system of measurement.

MEASURES OF AREA

Name	Sq. Met.	Sq. In.	Sq. Ft.	Acres
Sq. millimeter (mm.²)	= .0000010	= .001550 =		
Sq. centimeter (cm.²)	= .0001000	= .155003 =	.00107641	
Sq. decimeter (dm.²)	= .0100000	= 15.5003 =	.10764100	
Sq. meter or centare (m.² or ca.)	= 1.0000000	= 1,550.03 =	10.76410000	= .000247
Sq. decameter or are (Dm.² or A.)	= 100.0000000	= 155,003 =	1,076.4101	= .024710
Hectare	= 10,000.0000000	=	= 107,641.01	= 2.47110
Sq. kilometer	= .3861099 sq. mi.		= 10,764,101	= 247.110
Sq. myriameter	= 38.6109000 sq. mi.		=	= 24,711.0

MEASURES OF VOLUME

Name	Cu. Met.	Cu. In.	Cu. Ft.	Cu. Yd.
Cu. centimeter (cm.³)	= .000001	= .061025		
Cu. decimeter (dm.³)	= .001000	= 61.0254		
Centistere	= .010000	= 610.2540	= .35316	
Decistere	= .100000		= 3.53156	
Stere [=cu. m. (m.³)]	= 1.000000		= 35.3156	= 1.308
Decastere	= 10.000000		= 353.156	= 13.080

CURRENT MEASURES

LINEAL MEASURE

12	inches (in.)	= 1 foot	ft.
3	feet	= 1 yard	yd.
5.5	yards	= 1 rod	rd.
40	rods	= 1 furlong	fur.
8	furlongs	= 1 mile	mi.

In.	Ft.	Yd.	Rd.	Fur.	Mi
36 =	3 =				
198 =	16.5 =	5.5 =	1		
7,920 =	660 =	220 =	40 =	1	
63,360 =	5,280 =	1,760 =	320 =	8 =	1

Other units of measure are:

5 feet equal 1 pace.
2½ feet equal 1 military pace.
6 feet equal 1 fathom.
9 inches equal 1 span.
18 inches equal 1 cubit.
4 inches equal 1 hand (to measure horses).
21.8 inches equal 1 Bible cubit.

SURVEYOR'S MEASURE

7.92	inches	= 1 link	li.
25	links	= 1 rod	rd.

4 rods: } 100 links } = 1 chain ch. 66 feet: }

80	chains	= 1 mile	mi.

1 mi. = 80 ch. = 320 rd. = 8,000 li. = 63,360 in.

SQUARE MEASURE

144	square inches (sq. in.)	= 1 square foot	sq. ft.
9	square feet	= 1 square yard	sq. yd.
30¼	square yards	= 1 square rod	sq. rd.
160	square rods	= 1 acre	A.
640	acres	= 1 square mile	sq. mi.

Sq. mi.	A.	Sq. rd.	Sq. yd.	Sq. ft.	Sq. in.
1 =	640 =	102,400 =	3,097,600 =	27,878,400 =	4,014,489,600

SURVEYOR'S SQUARE MEASURE

625 square links (sq. li.). . . . . = 1 square rod. . . .sq. rd.
16 square rods . . . . = 1 square chain. .sq. ch.
10 square chains . . . . = 1 acre . . . . .A.
640 acres . . . . = 1 square mile. . .sq. mi.
36 sq miles (6 mi. square) . . = 1 township. . . . .Tp.
1 sq. mi. = 640 A. = 6,400 sq. ch = 102,400 sq. rd. = 64,000,000 sq. li.

The acre contains 4,840 square yards, or 43,560 square feet, and in form of a square is 208.71 feet on a side.

THE WEAR AND TEAR OF BUILDING MATERIALS

Material in Building	Frame dwelling		Brick dwelling (shingle roof)		Frame store		Brick store (shingle roof)	
	Average life Years	Per cent of depreciation per annum	Average life Years	Per cent of depreciation per annum	Average life Years	Per cent of depreciation per annum	Average life Years	Per cent of depreciation per annum
Brick.	—	—	75	1⅓	—	—	66	1½
Plastering.	20	5	30	3⅓	16	6	30	3⅓
Painting, outside	5	20	7	14	5	20	6	16
Painting, inside	7	14	7	14	5	20	6	16
Shingles	16	6	16	6	16	6	16	6
Cornice	40	2½	40	2½	30	3⅓	40	2½
Weather bo'ding	30	3⅓	—	—	30	3⅓	—	—
Sheathing.	50	2	50	2	40	2½	50	2
Flooring	20	5	20	5	13	8	13	8
Doors, complete.	30	3⅓	30	3⅓	25	4	30	3⅓
Windows, comp.	30	3⅓	30	3⅓	25	4	30	3⅓
Stairs and newel	30	3⅓	30	3⅓	20	5	20	5
Base	40	2½	40	2½	30	3⅓	30	3⅓
Inside blinds . . .	30	3⅓	30	3⅓	30	3⅓	30	3⅓
Building h'dware	20	5	20	5	13	8	13	8
Piazzas & porches	20	5	20	5	20	5	20	5
Outside blinds . .	16	6	16	6	16	6	16	6
Sills and first-floor joints . . .	25	4	40	2½	25	4	30	3⅓
Dimension lumbr	50	2	75	1⅓	40	2½	66	1½

These figures represent the averages deduced from the replies made by eighty-three competent builders unconnected with fire-insurance companies, in twenty-seven cities and towns of eleven Western States.

HOW TO FIGURE PLASTERING

Multiply the distance around the four sides of the room in feet by the height of the room in feet. Multiply the product by the price per square yard and divide this product by 9, because there are 9 square feet in a square yard. For the ceiling, multiply the length of the room by the width of the room in feet and then by the price per square yard, and divide by 9 as before. Add these two results and you have the entire cost of plastering the room.

To every barrel of lime estimate about ⅝ of a cubic yard of good sand for plastering.

One-third of a barrel of stucco will hard finish 100 square yards of plastering.

Six bushels of lime, 40 cubic feet of sand and 1½ bushels of hair will plaster 100 square yards with two coats of mortar.

In plastering, no deductions are made for openings, because it is considered that the extra work in finishing around them balances the material saved.

WEIGHTS OF PACIFIC COAST LUMBER

	Lbs. per M.
Oregon Fir, 1 inch, rough	2,200
Washington Red Cedar, 1 inch, rough	2,300
Washington Red Cedar, 1 inch, dressed	2,000
California Sugar Pine, 1 inch, rough	2,200
California Redwood, 1 to 2 inch, rough	2,500
California Redwood, 1 to 2 inch, S 1 S	2,200
California Redwood, 1 to 2 inch, S 2 S	2,000
Cedar Shingles, * A *	200

STANDARD WEIGHTS OF CYPRESS LUMBER

	Lbs. per M.
Lumber, rough, 2 inches and under	3,000
Lumber, rough, 2½ and 3 inches	3,500
⅞-inch Flooring and Ceiling	2,300

	Lbs. per M.
⅞-inch Ceiling	1,600
⅝-inch Ceiling	1,300
½-inch Ceiling	1,000
½-inch Bevel Siding	1,000
Shingles, all grades	300
⅜-inch Plaster Lath	500
⅝-inch Fence Lath	900
1¾ x 1¾ x 4 D. & H. Pickets	1,600
⅞ x 2½ x 4 D. & H. Pickets	1,800
2-inch O. G. Battens	500
2½-inch O. G. Battens	600
3-inch O. G. Battens	700

ESTIMATED WEIGHTS OF WHITE PINE

	Lbs. per M. Feet Green	Dry
Timbers, rough	3,250	2,500
Lumber, rough	3,000	2,400
Lumber, dressed	2,500	2,000
Lumber, D. & M.	2,400	1,800
Battens, O. G.	1,900	1,500
Siding and ⅝ Ceiling	1,250	800
Shingles	450	250
Lath	950	500

ESTIMATED WEIGHTS OF NORWAY PINE

	Lbs. per M. Feet Green	Dry
Timbers, rough	3,500	2,750
Lumber, rough	3,250	2,650
Lumber, dressed	2,900	2,300
Lumber, D. & M	2,600	2,000

These weights are taken from reports issued by the Agricultural Department of the United States.

ESTIMATING FRAME OR BALLOON BUILDINGS

In estimating the cost of labor necessary to convert rough lumber into available building material, the estimator should divide the labor as follows:

First, ascertain the cost of framing sills, joist, studs, rafters, and like dimension stuff on the ground ready to go into the building.

Second, estimate the cost of placing it on the building, or into the work. Siding, roof boards, sheathing, furring and flooring requires no primary labor to prepare it for the building; and, therefore, this class of material calls for the price of labor only to put it on the building.

The simplest method to estimate the labor of framing dimension or piece stuff, as scantling of all kinds, is by the thousand feet. A general rule adopted by us after a long experience and considerable investigation, is to add the entire bill of dimension stuff together, and price it for medium work at $4.00 per thousand for the labor of framing on the ground, and $5.00 per thousand for labor of working it into the building. We base our rule on the following demonstrations:

Two good carpenters will lay out and frame 50 pieces of 2 x 10 joist, 16 feet long, in a day of 9 hours, or about 1,350 feet; or they will frame 100 pieces of 2 x 6 studding, 12 feet long, in a day, or 1,200 feet; or they will frame 70 pieces of 2 x 6, 16 feet long, for rafters, in a day, or 1,120 feet; or they will frame 14 pieces of 8 x 8 sills, 16 feet long, or 1,190 feet. Calling carpenters' wages at $2.00 per day, we find that the framing of

Joist, 1,350 feet, cost	$5.00
Studding, 1,200 feet, cost	5.00
Rafters, 1,120 feet, cost	5.00
Sills, 1,190 feet, cost	5.00

Averaging the above, we find the price to be about $4.00 per 1,000 feet.

For siding, roof boards, sheathing and flooring, the price may be fixed as on the following basis:

Two good carpenters will put on 800 feet of lap siding in a day, or 1,600 feet of roof boards per day; staging not included. Calling wages at $3.00 per day, we find that to put on

Siding, costs $7.20 per 1,000 feet.
Roof boards, cost $3.80 per 1,000 feet.
Sheathing, costs $2.70 per 1,000 feet.

One good man will lay 900 feet of 1 × 6 matched flooring in a day, or 700 feet of 1 × 4 matched flooring in a day. At the same rate of wages the 1 × 6 floor will cost $3.25 per 1,000 feet to lay, and the 1 × 4 floor will cost $3.86 per 1,000 feet to lay.

A good man will carry up and lay on a roof from 1,600 to 2,400 shingles per day, which estimated at the same rate of wages and averaged, is $1.50 per 1,000.

Two men will put on 2,000 feet of felt paper per day, which being reduced from the same rate of wages, makes it cost 30 cents per square of 100 feet.

Two men will lay 500 to 600 feet of outside beaded ceiling work per day, or say $11.25 per 1,000 feet.

A man will put down 200 feet of plain base per day, or 100 feet of moulded base.

A man will fit and nail 400 pieces of bridging per day, or ¾ cent each.

Returning again to dimension stuff, as joist, studs, rafters, sills, etc., we find that two good men will place 50 pieces of 2 × 10 joist, 16 feet long, in a day, or 150 pieces of 2 × 6 studs, 12 feet long, in a day, or 100 pieces 2 × 6 rafters, 16 feet long, in a day, or 20 pieces of 8 × 8 sills in a day.

For the labor necessary to place material on a

building, some builders estimate labor by the square, as follows: Wages $3.00 per day.

Drop siding, 60 cents a square.
Lap siding, 72 cents a square.
Sheathing, 25 cents a square.
Surface boards, 30 cents a square.
Roof boards, plain, 30 cents a square.
Hip roofs, 60 cents a square.
Steep roofs, 65 cents a square.
Shingles, $1.10 to $1.25 a square.
Floor pine, 1 x 6, 35 to 60 cents a square.
Floor pine, 1 x 4, 35 to 60 cents a square.
Floor pine, 1 x 3, 75 cents to $1.25 a square.
Outside wall ceiling, $1.00 a square.
Soffit ceiling, $1.00 a square.
Wainscoting, from $2.00 to $3.00 a square.
Cleaning off pine floor, from 75 to 95 cents a square.

Tin work, valleys 14 inches wide, a man will lay from 1 to 1¼ square feet of valleys per day.

In closing this series of tables upon one of the most vital subjects connected with the building profession, I desire to call attention to the fact that the manner of taking out quantities in the United States is somewhat different from that of Europe, and especially that of England, where the rules and methods connected with this particular branch of building are settled and well defined. In the embryonic state of our building practice, we have no universal or general methods of drawing off quantities, excepting what has come out of necessty.

The time will doubtless come when we shall have a universal method that shall not only be thoroughly established by practice, but indorsed by the various building trades and architectural associations throughout the entire country, so that a mechanic, having become conversant with the rules and methods of New York, will not be called upon to study and make himself familiar with the rules and methods practiced in St. Louis or Chicago.

Large cities, by virtue of the facility for organization in the several branches of the building trades, are enabled to establish rules of measurement that govern their individual membership, but cannot control the conduct of other trades; hence, upon examination, it will be found that the rules of measurement for masonry in New York City vary from the rules in use in Cincinnati, Chicago and other large cities.

I am aware that the primary rules of mensuration, that is, the method of measuring any given surface or body, is governed by certain algebraic and mathematical calculations, which may be used by any one when he has mastered the proper method of procedure, and it is to illustrate and make plain this method that this book is written not only from a practical standpoint but from an American builder's view of the methods best adapted to the business interests of the builder.

Another somewhat different method than the foregoing is given herewith; it is taken from a trade journal of reliability, and possesses considerable merit. The system is all right, but the prices given are not to be followed, as they are much too low, not being within 25 to 35 per cent as high as current prices in the larger cities. This is especially arranged for balloon frame.

The first is an analysis of cost of four squares outside walls. For convenience, suppose a space 20 × 20 feet as a basis, resulting in 400 square feet, or 4 squares. The studding employed is 2 × 4 inch, sized on one side and one edge. The studding is placed 16 inches from centers and covered with dressed and matched stuff. Building paper is next laid on, and then first or second clear siding is used. Plates are included in the cost and are put on double thickness.

ANALYSIS OF OUTSIDE WALLS

19 pieces, 2 x 4 inch, 20 feet long = 247 feet, at $14.50 per M	$3.58
466 feet dressed and matched stuff, at $17.50	8.16
475 feet siding, at $21	9.97
11 pounds nails	.40
30 pounds paper, at 2½ cents per pound	.75
Framing and putting in place 247 feet of scantling, at $8 per M	1.98
Laying 4 squares of flooring, at 50 cents per square	2.00
Laying 4 squares of siding, at $1.12½ per square	4.50
Laying 4 squares, at 12½ cents per square	.50
Total	$31.84

Dividing this sum by 4 gives the price of a single square, $7.96.

The analysis of cost of 4 squares of roofing, the rafters being 2 = 4 inch scantling, set 2 feet between centers, covered with dressed and matched stuff, and the best quality of cedar shingles, laid 4½ inches to the weather, is as follows:

ANALYSIS OF ROOF WORK

12 scantlings, 2 x 4 inch, 20 feet long = 156 feet, at $14.50 per M	$2.26
466 feet matched stuff, at $17.50 per M	8.16
3⅓ M shingles, at $2.75 per M	9.17
14 pounds 3d. nails	.63
10 pounds 8d. and 10d. nails	.30
Framing and putting in place 156 feet 2 x 4 scantling, at $8 per M	1.25
4 squares of roof boarding, at 50 cents per square	2.00
4 squares of shingling, at $1.25 per square	5.00
Staging	.63
Total	**$29.40**

This sum, in turn, divided by 4 gives as the cost of a single square, $7.35.

The following is an analysis of cost of 4 squares of flooring, laid on joists 2 × 8 inches, the flooring being selected from No. 1 boarding, and the joists being placed 16 inches between centers. Allowance is made for doubling where necessary.

ANALYSIS OF FLOORING

17 joists, 2 x 8 inch, 20 feet long = 459 feet, at $14.50 per M	$6.65
466 feet of flooring, at $17.50 per M	8.15
15 feet of 1 x 2 inch bridging, at 2 cents	.30
10 pounds of 8d. common nails	.30
3 pounds of spikes	.08
Laying 4 squares of flooring, at 50 cents per square	2.00
Framing 459 feet of joists, at $5 per square	2.30
Bridging	.50
Total	$20.28

Dividing this amount by 4, as in the previous cases, gives $5.07 as the cost of 1 square of flooring.

The following is an analysis of the cost of an inside door, 2 feet 8 inches by 6 feet 10 inches, 1⅜ inches thick, cased and finished complete except the one item of painting:

ANALYSIS OF COST OF DOOR

Frame, 2 set casings and stops	$2.00
18 feet of moulding, at 2½ inches	.28
1 threshold, hardwood	.15
1 first quality door, size as given above	1.95
3½-inch morticed lock, bronze face, bolts and striking plate	.63
Porcelain knobs, plated roses and escutcheons	.40
1 pair of 3½-inch japan butts and screws	.25
Setting frame	.25
Casing up, 2 sides	.40
Putting down threshold	.15
Moulding, 1 side	.20
Fitting, hanging and trimming door	.75
Total	$7.42

The following is an analysis of cost of a 4-light window, with sash 14 × 30 inches, 1⅜ inches thick, check-rail, the window set, cased and finished complete.

ANALYSIS OF COST OF WINDOW

Window frame prepared for weights	$2.15
Sash glazed	2.10
20 feet 2½-inch moulding	.30
25 feet inside case and window sill	.75
28 pounds of sash weights	.56
Sash cord	.18
Grounds for plastering and putting on	.30
Setting frame	.25
Casing up	.55
Fitting sash	.15
Nails	.10
Sash lock	.25
Putting on sash lock	.10
Total	$7.64

Add to the foregoing not less than 30 per cent, but it is better in all cases that local prices of material and labor be embodied in the analysis.

ESTIMATING FOR OUTSIDE DOOR AND WINDOW FRAMES

For ordinary buildings, either wood or brick, the following prices, which are for labor only, will be found to be as nearly correct as possible where local conditions are unknown. For simply making the frames, setting same, hanging sashes, doors, blinds, etc., the number that can be made, hung, or set in a day of nine hours, is given, as well as the price which will enable the estimator to tell approximately the cost of any number of frames either in place or out.

	No. of Pieces in Day's Work	Price for Each
Making plane frames for weights	3	$1.00
Setting frames in wall	14	.22
Hanging outside blinds	10	.30
Hanging inside blinds, 50c. to $1.00	5	.60
Fitting sash per window	18	.18
Hanging sash, trimming, locks and lifts	14	.23
Casing	10	.30
Putting on stops	35	.09
Band moulding	25	.12
Fitting stool	13	.24
Fitting apron	25	.12
Total		$3.40

Fitting and hanging doors on outside frames, trimming with 4-inch loose pine, joint hinges, mortise lock, bronze or plated rose, hardwood knob, night latch, and all complete, three hinges to the door, door 1¾-inch thick, pine, to complete $1.95. If two hinges, and 1⅜-inch door, $1.50. If hardwood, add 15 per cent.

If frames are bought at the factory all ready made, no blinds to hang, no band mouldings to plant, then the cost for setting, hanging, casing complete on one side, doors or windows, will be $1.25.

The average quantity of material required to make frames for common houses, running measure, allowing for waste and joints on the basis of a 2-light window, with glass 24 × 36 inches, and a door measuring 2 feet 8 inches by 6 feet 8 inches, is given in the following table, which covers all the items required to complete common frames:

	Feet
Window jambs and heads, with drip on sill	18
Door jambs and heads	18
Outside casing, window	18
Outside casing, door	19
Inside casing, window, with apron	20
Inside casing, door, each side	18

About the same number of feet in length will be required for mouldings and stops.

TABLE FOR ESTIMATING NAILS

1000 shingles require 3½ pounds 4d. nails.
1000 lath require 6½ pounds 3d. nails.
1000 feet of beveled siding require 18 pounds 6d. nails.
1000 feet of sheeting requires 20 pounds 8d. nails.
1000 feet of sheeting requires 25 pounds 10d. nails.
1000 feet of flooring requires 30 pounds 8d. nails.
1000 feet of flooring requires 35 pounds 10d. nails.
1000 feet of studding requires 14 pounds 10d. nails.
1000 feet of studding requires 10 pounds 20d. nails.
1000 feet of furring, 1 x 2, requires 10 pounds 10d. nails.
1000 feet of ⅞ finish requires 30 pounds of 8d. nails.
1000 feet of 1⅛ finish requires 40 pounds 10d. finish nails.

The following table shows the name, length and number of nails to the pound of the different sizes:

NUMBER OF NAILS TO THE POUND

Name	Length	No. to a pound
3d fine	1 inch	1150
3d common	1¼ inch	720
4d common	1⅜ inch	432
5d common	1½ to 1¾ inch	352
6d finish	2 inch	350
6d common	2 inch	252
7d common	2½ inch	192
8d finish	2½ inch	190
8d common	2½ inch	132
9d common	2¾ inch	110
10d finish	3 inch	137
10d common	3 inch	87
12d common	3¼ inch	66
20d common	3⅝ inch	35
30d common	4 inch	27
40d common	4½ inch	21
50d common	5⅛ inch	15

PAINTERS' MEASUREMENTS

In England the custom is to employ a clerk quick at figures, whose duty it is to take off, from the plans and specifications, an accurate list of all the materials and labor required in the performance of the work, setting down in each case the number of yards or feet, as the case may be, of each item. In the case of painting, the figures obtained for the carpenter and joiner prove of service also for the work to be done by the painter. The following is a table that is intended to indicate the method of measurement of painters' work, and also the order in which the various items may be taken. A similar table added to, or changed, as might be necessary to suit American methods of construction, would be very useful to have on hand when getting out estimates, as it would insure nothing being left out. The table which follows accurately indicates the English practice.

Lead, in oil on white work, at——per yard super.
" " cement " " " " "
Ornamental railings, etc., " " " "
Skylights, " " " "
Skirtings, 12-in. girth and under, at——per foot run.
Strings, " " " " " " "
Chair rails, " " " " " " "
Hand " " " " " " " "
Balusters, " " " " " "
Newels, " " " " " "
Rain pipes, " " " " " "
Ornamental heads " number.
Ears, " "
Shoes, " "
Eaves, gutter " foot run.
Stopped ends, " number.
Outlets, " "
Swan necks, " "

Cement reveals (jambs)	at	foot run.
Cornices under — girth,	"	"
Window sills, "	"	"
Coping edge, "	"	"
Stone strings, "	"	"
Stone plinths, "	"	"
Iron castings, "	"	"
Grate bars, "	"	"
Sash squares,	"	dozen.
Sash frames,	"	number.
Small "	"	"
Two-light casement frames,	"	"
Four " " "	"	"
Sash squares,	"	dozen.
Brackets,	"	number.
Finials,	"	"
Step ladders,	"	"
Dressers,	"	"
Chimney pieces,	"	"
Four oils and extra finished varnish, gray,	"	Yard super.
Grainer; extra grain for wainscot and twice varnish,	"	" "
Grainer; extra grain enrichment for brackets 4 in. wide,	"	foot run.
Stainer; stain to an approved tint and twice varnish with the best copal varnish,	"	foot super.
French polisher; French polishing,	"	" "
French polishing to hand rails,	"	foot run.
Gilder; gilding on flat surface,	"	foot super.
Gilder on carved work, stating height and description,	"	foot run.
Moulded work, stating girth,	"	" "
Boards, etc., " "	"	" "
Carved caps,	"	" "
Brasses and simple items of a similar nature,	"	" "

TO FIND THE NUMBER OF SINGLE ROLLS OF PAPER NEEDED FOR ANY GIVEN ROOM.

To find the number of single rolls required for a wall, multiply the distance around the room by the height, taking out 20 square feet for each opening, and divide by 30. To find the number of rolls for the ceiling, multiply the length by the width and divide by 30. The number of yards of border required can easily be measured.

For example, room 12 x 14, 10 feet high, two doors and three windows:

Length, two walls, 14 feet each	28 feet
Width, two walls, 12 feet each	24 "
	52 "
Multiply by height	10 "
	520 "
Less five openings, allowing 20 sq. ft. for each ..	100 "
Divided by number of sq. ft. in a roll	30)420(14 rolls required
	30
	120
	120

To find the quantity of border required, divide length around the room, 52 feet, by 3, equal to about 18 yards.

The price of border is for a single strip, the width of the border and one yard long.

The price of the paper is for a single roll, one-half yard wide and eight yards long. Allowing for all waste, this will cover 30 square feet.

The following table will be useful to the estimator:

Per 1000 Feet	Per Foot Cube	1	2	3	4	5	6	7	8	9	10	11	12	13	14	15	16	17	18
$20.00	.24	⅛	¼	½	⅝	⅞	1	1⅛	1¼	1½	1⅝	1⅞	2	2⅛	2¼	2½	2⅝	2⅞	3
22.50	.27	¼	⅜	⅝	¾	1	1⅛	1¼	1½	1⅝	1⅞	2	2¼	2½	2⅝	2⅞	3	3⅛	3⅜
25.00	.30	¼	⅜	⅝	¾	1	1¼	1⅜	1⅝	1⅞	2	2¼	2½	2¾	2⅞	3⅛	3¼	3½	3¾
27.50	.33	¼	½	¾	1	1⅛	1⅜	1½	1⅞	2	2¼	2½	2¾	3	3¼	3½	3⅝	4	4⅛
30.00	.36	¼	½	¾	1	1¼	1½	1¾	2	2¼	2½	2¾	3	3¼	3½	3¾	4	4¼	4½
32.50	.39	¼	½	⅞	1⅛	1⅜	1⅝	1⅞	2⅛	2½	2¾	3	3¼	3½	3⅞	4	4⅜	4½	4⅞
35.00	.42	¼	⅝	⅞	1¼	1½	1¾	2	2¼	2⅝	3	3¼	3½	3¾	4	4⅜	4⅝	5	5¼
37.50	.45	⅜	⅝	1	1¼	1½	1⅞	2¼	2½	2¾	3⅛	3½	3¾	4	4⅜	4¾	5	5¼	5⅝
40.00	.48	⅓	⅔	1	1⅓	1⅔	2	2⅓	2⅔	3	3⅓	3⅔	4	4⅓	4⅔	5	5⅓	5⅔	6
45.00	.54	⅜	¾	1⅛	1½	1⅞	2¼	2⅝	3	3⅜	3¾	4⅛	4½	4⅞	5¼	5⅝	6	6⅜	6¾
50.00	.60	½	⅞	1¼	1⅔	2	2½	3	3⅜	3¾	4	4½	5	5⅜	5¾	6¼	6⅝	7	7½
55.00	.66	½	1	1⅜	1⅞	2¼	2¾	3¼	3¾	4⅛	4⅝	5	5½	6	6⅜	6⅞	7	8	8¼
60.00	.72	½	1	1½	2	2½	3	3½	4	4½	5	5½	6	6½	7	7½	8	8½	9
80.00	.96	⅔	1⅓	2	2⅔	3⅓	4	4⅔	5⅓	6	6⅔	7⅓	8	8⅔	9⅓	10	10⅔	11⅓	12
100.00	1.20	⅞	1⅔	2½	3⅓	4	5	6	6⅔	7½	8	9	10	11⅛	11⅞	12½	13⅓	14	15

Per 1000 Feet	19	20	21	22	23	24	25	26	27	28	29	30	31	32	33	34	35	36
$20.00	3⅛	3¼	3½	3⅝	3⅞	4	4⅛	4¼	4½	4⅝	4⅞	5	5⅛	5¼	6½	5⅝	5⅞	6
22.50	3½	3¾	4	4⅛	4⅜	4½	4¾	4⅞	5⅛	5¼	5½	5⅝	5¾	6	6¼	6⅜	6½	6¾
25.00	4	4⅛	4⅜	4½	4¾	5	5¼	5⅜	5⅝	5¾	6	6¼	6½	6⅝	6⅞	7	7¼	7½
27.50	4¼	4½	4¾	5	5¼	5½	5¾	6	6¼	6½	6⅝	6⅞	7⅛	7⅜	7½	7¾	8	8¼
30.00	4¾	5	5¼	5½	5¾	6	6¼	6½	6¾	7	7¼	7½	7¾	8	8¼	8½	8¾	9
32.50	5⅛	5½	5¾	6	6¼	6½	6¾	7	7¼	7½	7¾	8⅛	8½	8¾	9	9¼	9½	9¾
35.00	5½	5¾	6⅛	6⅜	6¾	7	7¼	7½	7⅞	8⅛	8½	8¾	9	9⅜	9⅝	10	10¼	10½
37.50	6	6¼	6½	7	7¼	7½	7¾	8	8⅜	8¾	9	9⅜	9¾	10	10⅜	10⅝	11	11¼
40.00	6⅓	6⅔	7	7⅓	7⅔	8	8⅓	8⅔	9	9⅓	9⅔	10	10⅓	10⅔	11	11⅓	11⅔	12
45.00	7⅛	7½	7⅞	8¼	8⅝	9	9⅜	9¾	10⅛	10½	10⅞	11¼	11⅝	12	12⅜	12¾	13⅛	13½
50.00	8	8⅜	8¾	9⅛	9½	10	10⅜	10¾	11¼	11½	12	12½	13⅛	13½	13¾	14	14½	15
55.00	8½	9	9½	10	10½	11	11½	12	12⅜	12⅞	13⅜	13¾	14¼	14⅝	15	15½	16	16½
60.00	9½	10	10½	11	11½	12	12½	13	13½	14	14½	15	15½	16	16½	17	17½	18
80.00	12⅔	13⅓	14	14⅔	15⅓	16	16⅔	17⅓	18	18⅔	19⅓	20	20⅔	21⅓	22	22⅔	23⅓	24
100.00	16	16⅔	17½	18⅓	19	20	20¾	21½	22½	23	24	25	26	26⅔	27½	28	29	30

EXAMPLE 1.—Required, cost of 3x11-inch timber per foot run, at the rate of $32.50 per thousand feet.
Rule.—Find area figures on top line—33. Under this and in line with $32.50 is figure 9. Answer, 9 cents per foot.

EXAMPLE 2.—Required, price of 4x12-inch stone per foot run when cost per foot cube is 30 cents.
(Note.—As table does not run so high, find, say, half of area—24.)
Under 24 and in line with 30 is 5; twice this equals 10 cents—answer.
To find odd areas, or higher and lower figures, multiply, divide or add as required.

CONCLUSION

In conclusion I would suggest the following simple method of keeping a record of cubic contents and cost, and would say that the information an architect has of this kind from his own buildings is the best for him, as it is probable that no other architect is quite similar in his style of work and finish.

A book or a number of sheets of paper should be ruled in suitable widths for the following columns: 1st, date (year); 2d, name of building (for owner); 3d, where erected; 4th, short description; 5th, cubical contents in feet; 6th, cost of building; 7th, cost per cubic foot; 8th, remarks. The kinds of buildings should be classified so that prices of one class may be seen and compared at a glance in one column. An example is here shown.

Date	Name	Where Built	Description	Cubic Feet	Cost $ c.	Cts. per c. ft.	Remarks
......	....	....	....	......	.... ..	....	....
......	....	....	....	......	.... ..	....	....
......	....	....	....	......	.... ..	....	....
......	....	....	....	......	.... ..	....	....

In computing the cubical contents the rule most commonly used is to measure the building as a whole or in parts from the bottom of the footings to a point halfway up the slope of roof, this being done in parts where there are different heights of roofs, towers, etc. In measuring brick or stone buildings, light wooden porticos or verandas are usually omitted. There should be a uniform system of omitting or including such items as heating, mantels, grates and tiles, electric wiring, or of noting two rates, one omitting and the other including these.

Indeed, an exact record of the cost of all buildings the contractor may erect, should be kept, and anything peculiar or uncommon or unusual should be noted, that in the future the knowledge obtained in this manner may be put to good purpose.

In this work I have described several schemes for estimating and have given my views as to their respective merits, and the more I have examined into the question of estimating, the more I am confirmed in the views expressed in the first pages of the work, namely, that "no exact methods of estimating can be given," and that the best and most reliable way is to estimate in detail. All other methods have certain good points, but, as a rule, they lack reliability, a quality the young contractor does not want to be up against; so it is better he should follow the safer, if more laborious way of figuring on every item going into the building he is about to tender for. A celebrated artist once explained that his success as a painter arose from his following the rule, "First know what you want to do, and then do it." So here, before anything can be done, it is necessary for careful plans to be made to show what is wanted, and these plans should be carefully studied and every item shown in them or described in the specifications should be noted.

Trusting my efforts will prove useful to the young and progressive workman who has a desire to become a contractor, and that they will aid and assist him in bettering himself, and with this hope in view I close this volume.

INDEX

HOUSE PLAN SUPPLEMENT

PERSPECTIVE VIEWS AND FLOOR PLANS

of Fifty Low *and* Medium Priced Houses

FULL AND COMPLETE WORKING PLANS AND SPECIFICATIONS OF ANY OF THESE HOUSES WILL BE MAILED AT THE LOW PRICES NAMED, ON THE SAME DAY THE ORDER IS RECEIVED.

Other Plans

WE ILLUSTRATE IN ALL BOOKS UNDER THE AUTHORSHIP OF FRED T. HODGSON FROM 25 TO 50 PLANS, NONE OF WHICH ARE DUPLICATES OF THOSE ILLUSTRATED HEREIN.

FOR FURTHER INFORMATION, ADDRESS THE PUBLISHERS.

SEND ALL ORDERS FOR PLANS TO

FREDERICK J. DRAKE & COMPANY

ARCHITECTURAL DEPARTMENT

211-213 East Madison Street, CHICAGO

"The Mokahi"

Price of Plans and Specifications

$5.00

Full and complete plans and specifications of this house will be furnished for $5.00.
Cost of this house is about $1,650.

Floor Plans of "The Mokahi"

First Floor Plan.

SIZE:
Width, 20 feet
Length, 44 feet

Blue prints consist of cellar and foundation plan; roof plan; floor plan; front and side elevations.

Complete typewritten specifications with each set of plans.

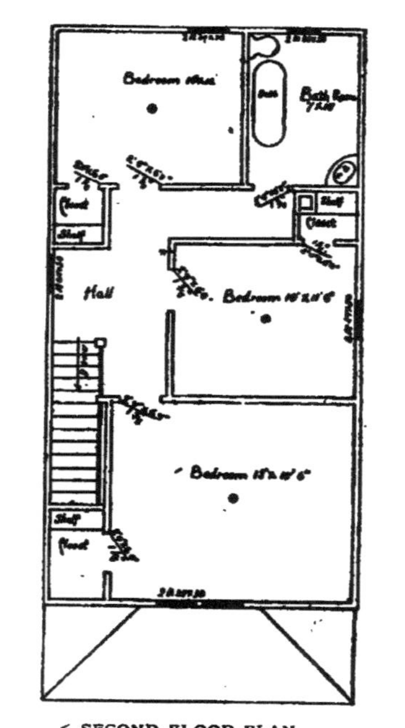

SECOND FLOOR PLAN

"The Dionelli"

Price of Plans and Specifications

$5.00

Full and complete plans and specifications of this house will be furnished for $5.00.
Cost of this house is from $2,500 to $2,700, according to the locality in which it is built.

Floor Plans of "The Dionelli"

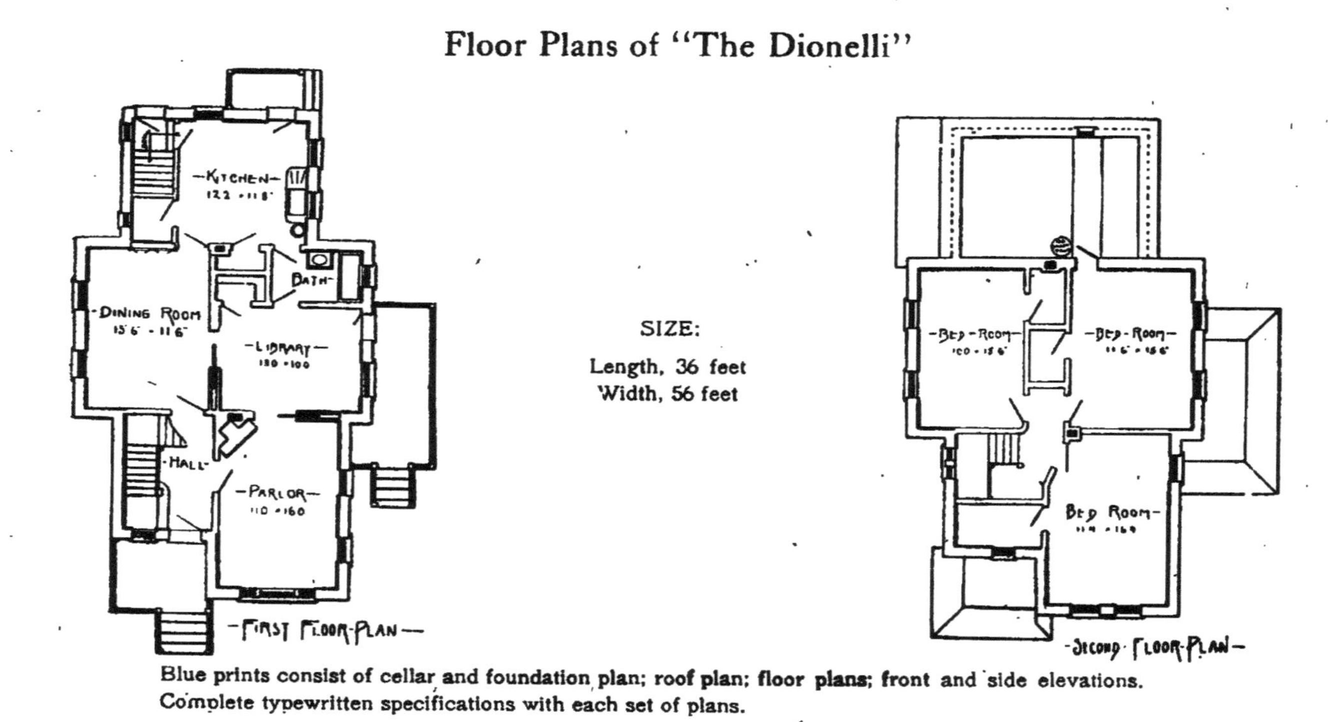

SIZE:

Length, 36 feet
Width, 56 feet

Blue prints consist of cellar and foundation plan; roof plan; floor plans; front and side elevations. Complete typewritten specifications with each set of plans.

"The Adele"

Price of Plans and Specifications

$5.00

Full and complete plans and specifications of this house will be furnished for $5.00.

Cost of this house is about $2,400.

Floor Plans of "The Adele"

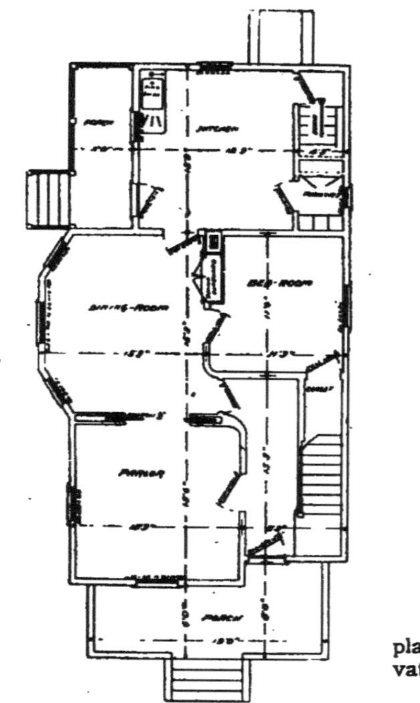

FIRST FLOOR PLAN

SIZE:
Width, 24 feet
Length, 56 feet

Blue prints consist of cellar and foundation plan; roof plan; floor plan; front and side elevations.

Complete typewritten specifications with each set of plans.

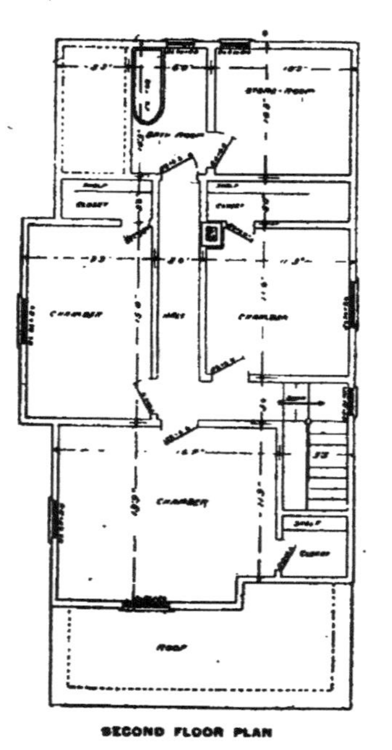

SECOND FLOOR PLAN

"The Parisian"

Price of Plans and Specifications

$5.00

Full and complete plans and specifications of this house will be furnished for $5.00.
Cost of this house is from $2,100 to $2,200, according to the locality in which it is built.

Floor Plan of "The Parisian"

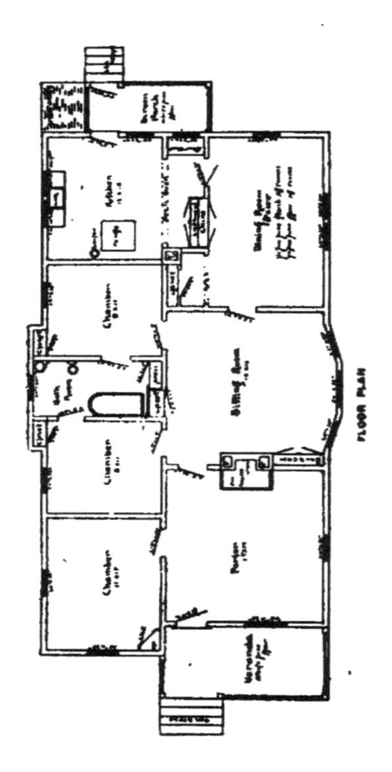

SIZE
Width, 29 feet
Length, 62 feet

Blue prints consist of cellar and foundation plan; roof plan; floor plan; front and side elevations.

Complete typewritten specifications with each set of plans.

"The Feeney"

Price of Plans and Specifications

$5.00

Full and complete plans and specifications of this house will be furnished for $5.00.
Cost of this house is from $1,000 to $1,100, according to the locality in which it is built.

Floor Plans of "The Feeney"

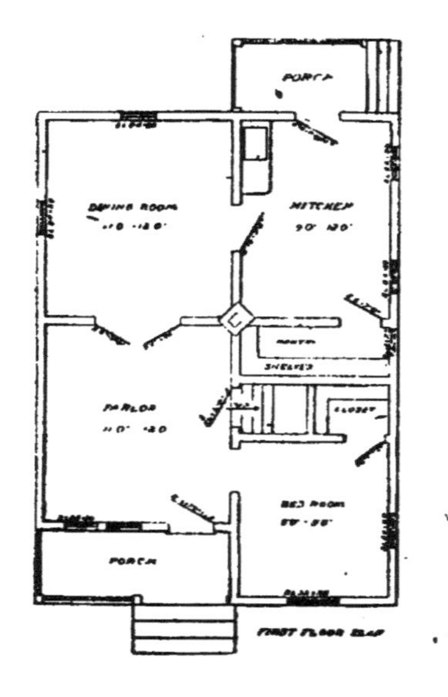

SIZE:

Length, 22 feet
Width, 32 feet

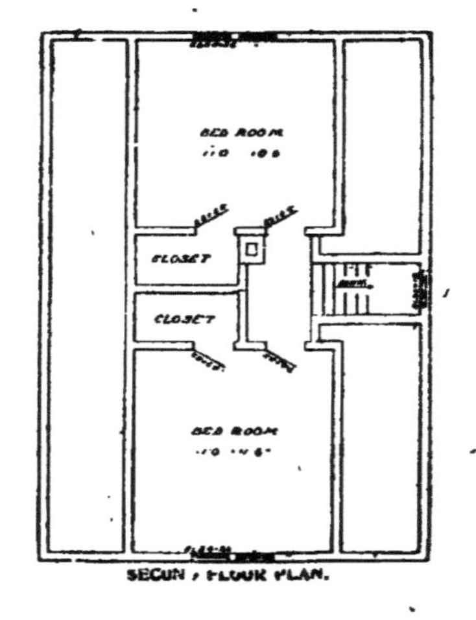

Blue prints consist of cellar and foundation plan; roof plan; floor plans; front and side elevations. Complete typewritten specifications with each set of plans.

"The Peri"

Price of Plans and Specifications

$5.00

Full and complete plans and specifications of this house will be furnished for $5.00.
Cost of this house is about $2,100.

Floor Plans of "The Peri"

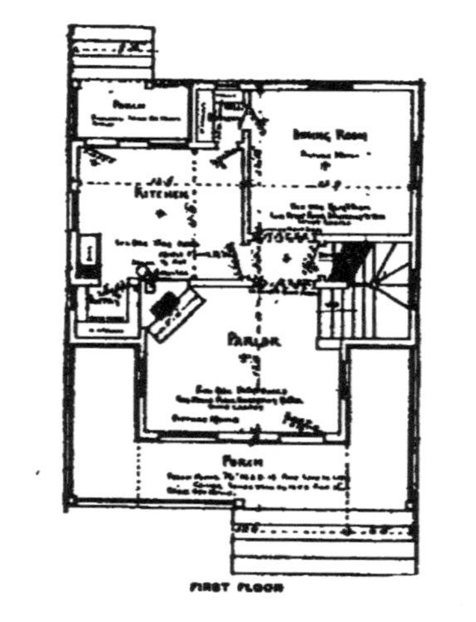

SIZE:

Length, 28 feet
Width, 46 feet

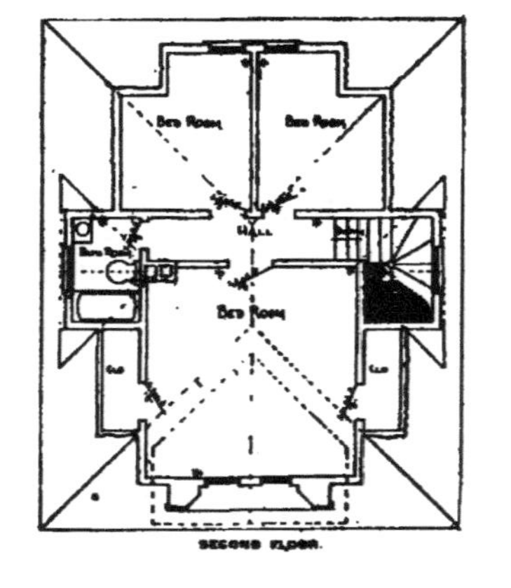

Blue prints consist of cellar and foundation plan; roof plan; floor plans; front and side elevations. Complete typewritten specifications with each set of plans.

"The Oklahoma"

Price of Plans and Specifications

$5.00

Full and complete plans and specifications of this house will be furnished for $5.00.

Cost of this house is from $1,200 to $1,300, according to the locality in which it is built.

Floor Plan of "The Oklahoma"

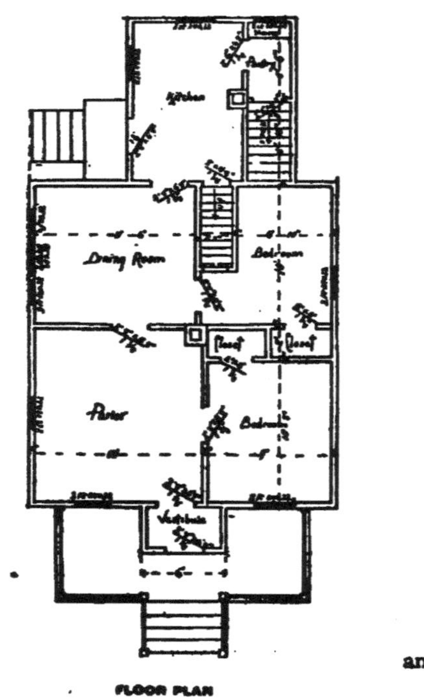

FLOOR PLAN

SIZE:
Width, 20 feet
Length, 44 feet

Blue Prints consist of cellar and foundation plan; roof plan; floor plan; front and side elevations.

Complete typewritten specifications with each set of plans.

"The Macatawa"

Price of Plans and Specifications

$5.00

Full and complete plans and specifications of this house will be furnished for $5.00.
Cost of this house is from $1,350 to $1,400, according to the locality in which it is built.

Floor Plans of "The Macatawa"

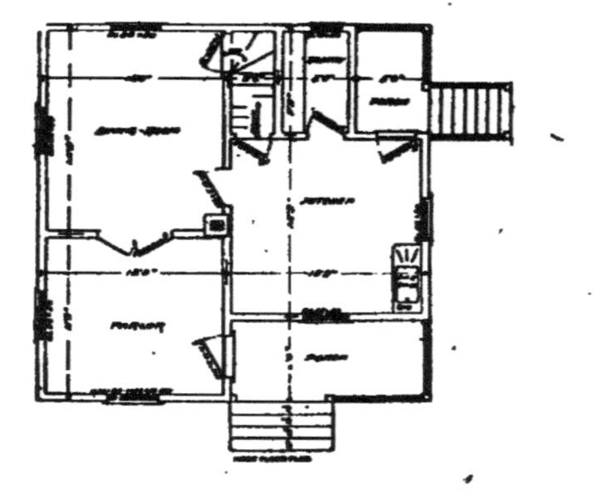

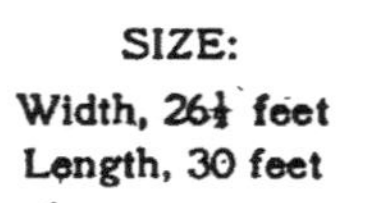

SIZE:
Width, 26½ feet
Length, 30 feet

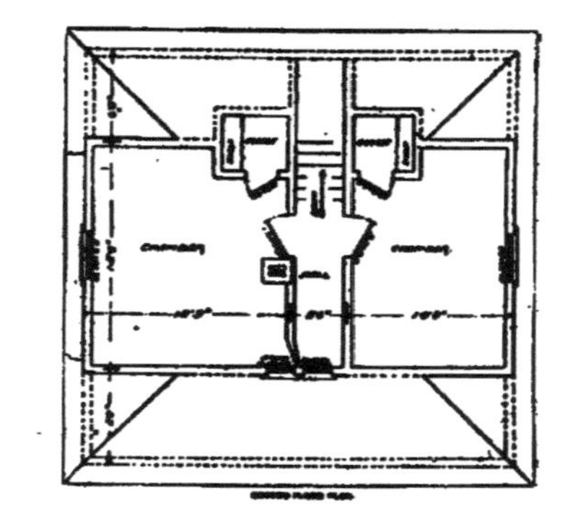

Blue prints consist of cellar and foundation plan; roof plan; floor plan; front and side elevations. Complete typewritten specifications with each set of plans.

"The American"

Price of Plans and Specifications

$5.00

Full and complete working plans and specifications of this house will be furnished for $5.00.
Cost of this house is about $500, according to the locality in which it is built.

Floor Plan of "The American"

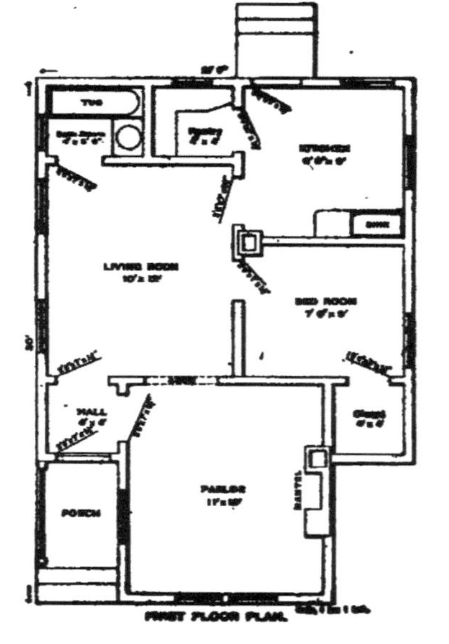

SIZE

Width, 22 feet
Length, 30 feet

Blue prints consist of foundation plan; floor plan; roof plan; front and side elevations. Complete typewritten specifications with each set of plans.

"The Pomeroy"

Price of Plans and Specifications

$5.00

Full and complete working plans and specifications of this house will be furnished for $5.00. Cost of this house is about $1,000, according to the locality in which it is built.

Floor Plan of "The Pomeroy"

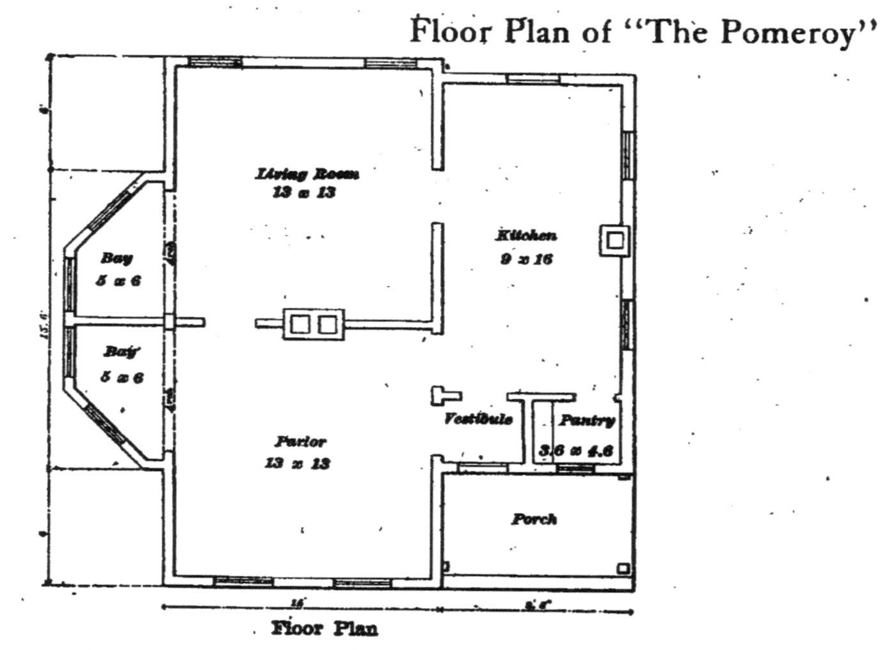

Floor Plan

SIZE
Width, 29 feet
Length, 27 feet

Blue prints consist of foundation plan; floor plan; roof plan; front and side elevations. Complete typewritten specifications with each set of plans.

"The Gastonia"

Price of Plans and Specifications

$5.00

Full and complete working plans and specifications of this house will be furnished for $5.00. Cost of this house is from $1,250 to $1,300, according to the locality in which it is built.

Floor Plan of "The Gastonia"

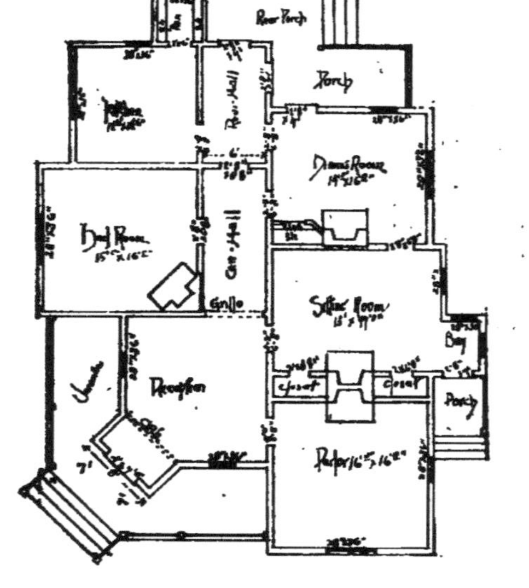

SIZE

Width, 30 feet
Length, 42 feet

Blue prints consist of floor plan; roof plan; front and side elevations.
Complete typewritten specifications with each set of plans.

"The Ontario"

Price of Plans and Specifications

$6.00

Full and complete working plans and specifications of this house will be furnished for $6.00. Cost of this house is from $3,000 to $3,100, according to the locality in which it is built.

FIRST FLOOR PLAN

Floor Plans of "The Ontario"

SIZE

Width, 23 feet
Length, 50 feet

Blue prints consist of cellar and foundation plan; floor plans; roof plan; front and side elevations.

Complete typewritten specifications with each set of plans.

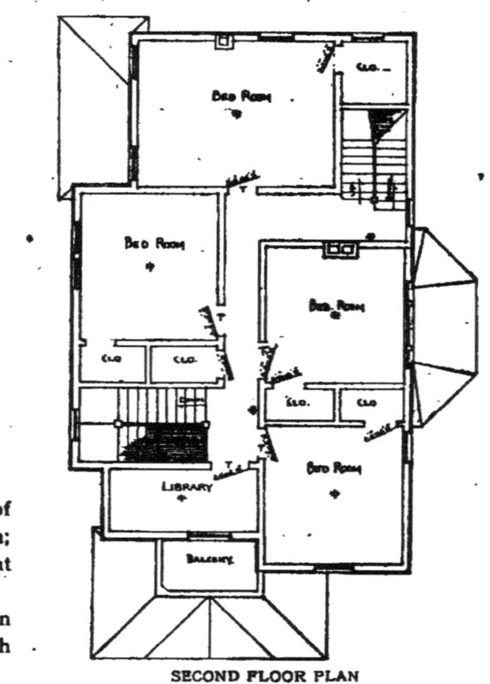

SECOND FLOOR PLAN

"The Eastlake"

Price of Plans and Specifications

$5.00

Full and complete working plans and specifications of this house will be furnished for $5.00.
Cost of this house is from $1,900 to $2,000, according to locality in which it is built.

Floor Plans of "The Eastlake"

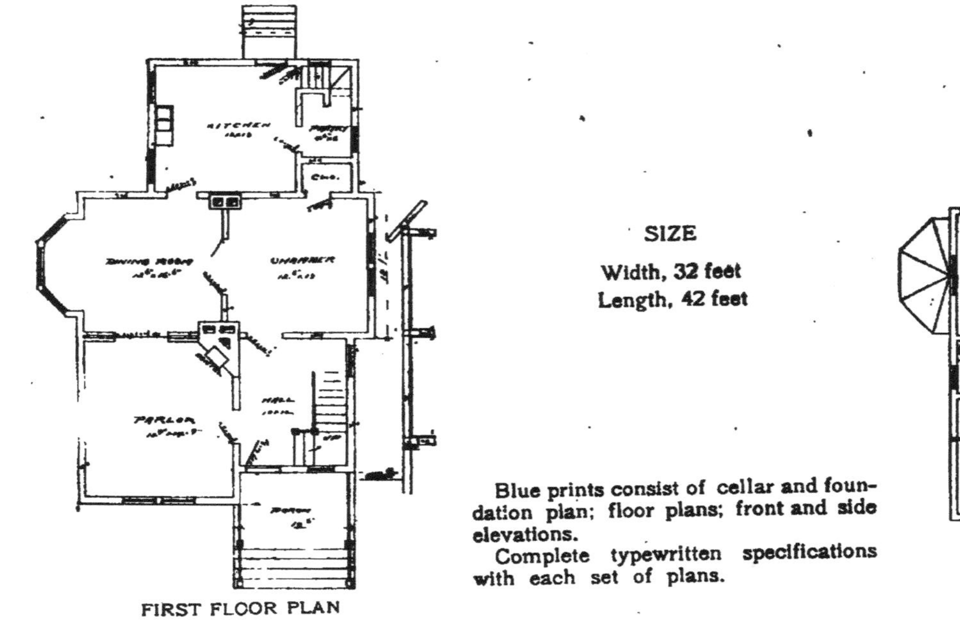

FIRST FLOOR PLAN

SIZE

Width, 32 feet
Length, 42 feet

Blue prints consist of cellar and foundation plan; floor plans; front and side elevations.

Complete typewritten specifications with each set of plans.

SECOND FLOOR PLAN

"The Shaw"

Price of Plans and Specifications

$8.00

Full and complete working plans and specifications of this house will be furnished for $8.00.
Cost of this house is from $4.200 to $4,400, according to the locality in which it is built.

Floor Plans of "The Shaw"

SIZE

Width, 32 feet
Length, 54 feet

FIRST FLOOR PLAN

SECOND FLOOR PLAN

Blue prints consist of cellar and foundation plan; floor plans; roof plan; front and side elevations. Complete typewritten specifications with each set of plans.

"The Egan"

Price of Plans and Specifications

$5.00

Full and complete working plans and specifications of this house will be furnished for $5.00. Cost of this house is from $2,200 to $2,400, according to the locality in which it is built.

Floor Plans of "The Egan"

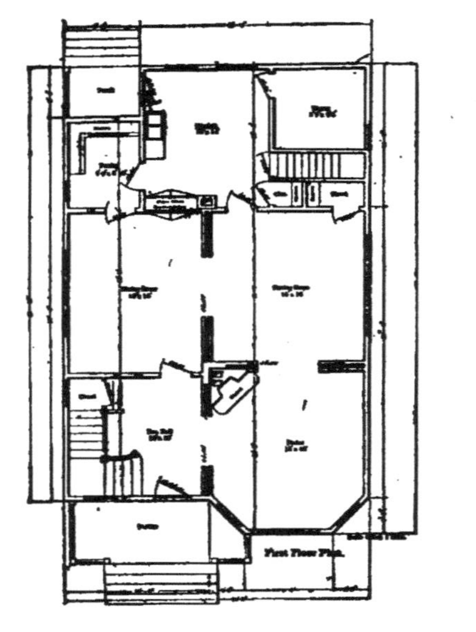

SIZE

Width, 30 feet
Length, 48 feet

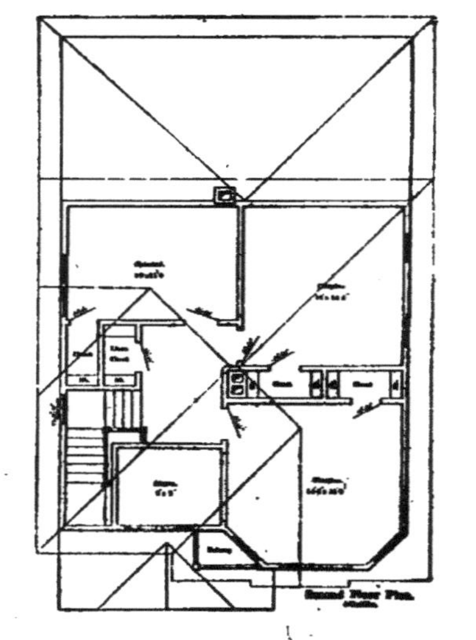

Blue prints consist of cellar and foundation plan; roof plan; floor plans; front and side elevations. Complete typewritten specifications with each set of plans.

"The Southern"

Price of Plans and Specifications

$5.00

Full and complete working plans and specifications of this house will be furnished for $5.00. Cost of this house is from $1,700 to $1,800, according to the locality in which it is built. Special itemized estimate of cost for $1.00 extra.

Floor Plans of "The Southern"

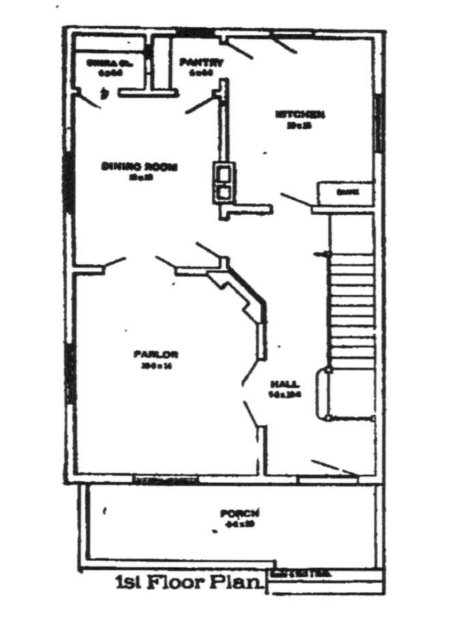

1st Floor Plan.

SIZE

Width, 22 feet
Length, 32 feet
Exclusive of Porch

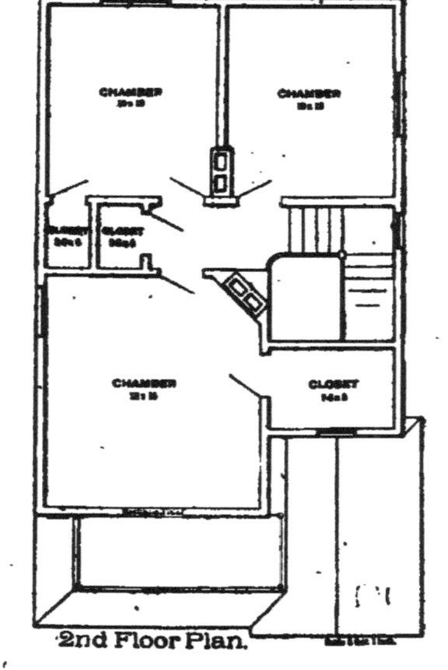

2nd Floor Plan.

Blue prints consist of cellar and foundation plan; floor plans; roof plan; front and side elevations. Complete typewritten specifications with each set of plans.

"The Minnetonka"

Price of Plans and Specifications

$5.00

Full and complete working plans and specifications of this house will be furnished for $5.00.
Cost of this house is from $1,700 to $1,800, according to the locality in which it is built.

Floor Plans of "The Minnetonka"

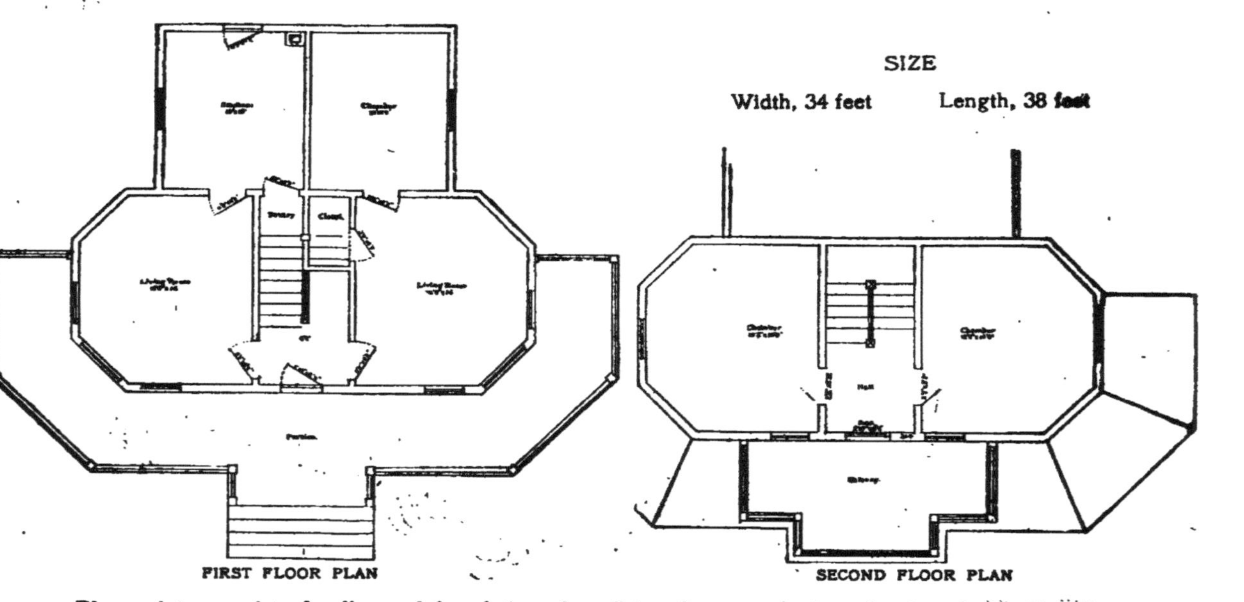

Blue prints consist of cellar and foundation plan; floor plans; roof plans; front and side elevations. Complete typewritten specifications with each set of plans.

"The Linwood"

Price of Plans and Specifications

$5.00

Full and complete working plans and specifications of this house will be furnished for $5.00. Cost of this house is from $600 to $750, according to the locality in which it is built.

Floor Plan of "The Linwood"

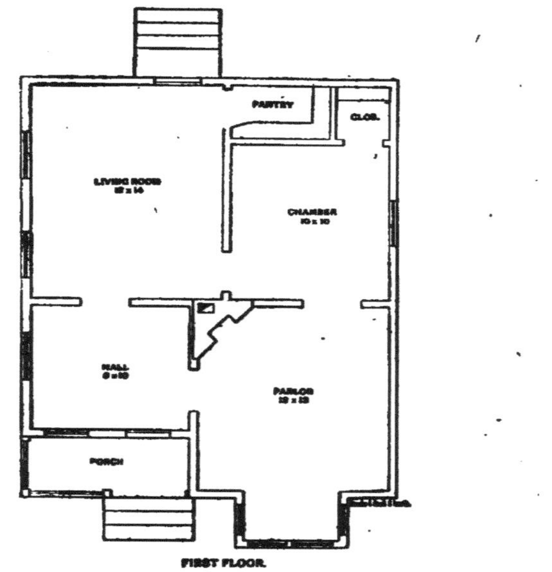

Blue prints consist of floor plan; roof plan; front and side elevations.
Complete typewritten specifications with each set of plans.

SIZE

Width, 24 feet
Length, 32 feet

"The St. Charles"

Price of Plans and Specifications

$5.00

Full and complete working plans and specifications of this house will be furnished for $5.00. Cost of this house is from $1,400 to $1,500, according to the locality in which it is built.

Floor Plans of "The St. Charles"

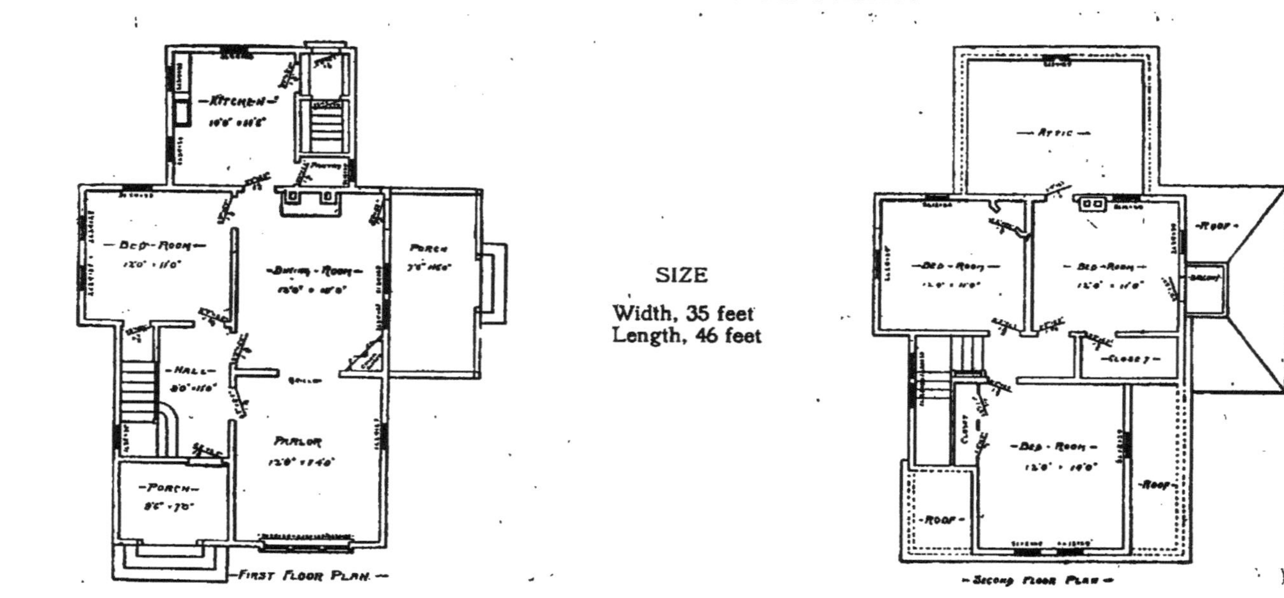

SIZE

Width, 35 feet
Length, 46 feet

Blue prints consist of cellar and foundation plan; roof plan; floor plans; front and side elevations. Complete typewritten specifications with each set of plans.

"The Newark"

Price of Plans and Specifications

$10.00

Full and complete working plans and specifications of this house will be furnished for $10.00.
Cost of this house is from $4,900 to $5,000, according to the locality in which it is built.

Floor Plans of "The Newark"

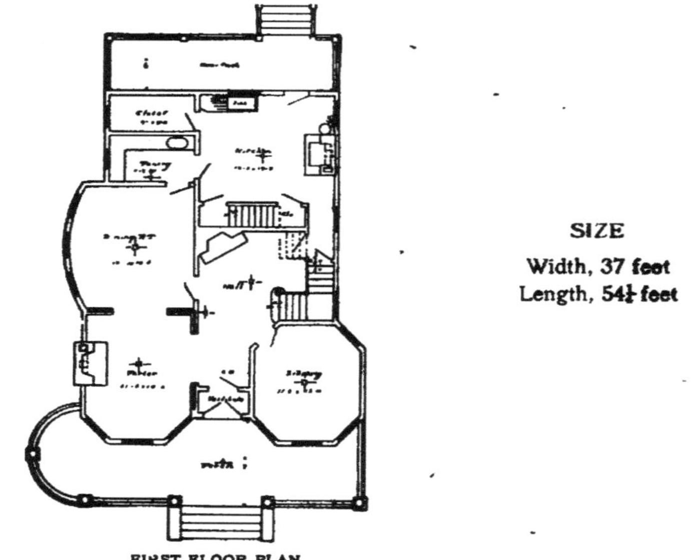

FIRST FLOOR PLAN

SIZE

Width, 37 feet
Length, 54½ feet

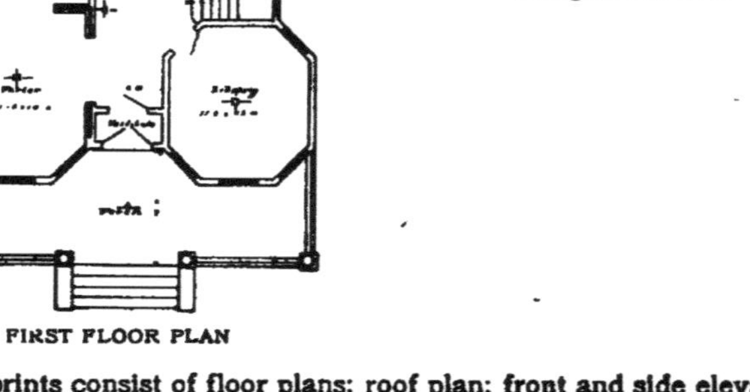

SECOND FLOOR PLAN

Blue prints consist of floor plans; roof plan; front and side elevations.
Complete typewritten specifications with each set of plans.

"Country School House"

Price of Plans and Specifications

$5.00

Full and complete working plans of this school house will be furnished for $5.00. This school building has been erected at a cost of $1,600.

Floor Plan of a "Country School House"

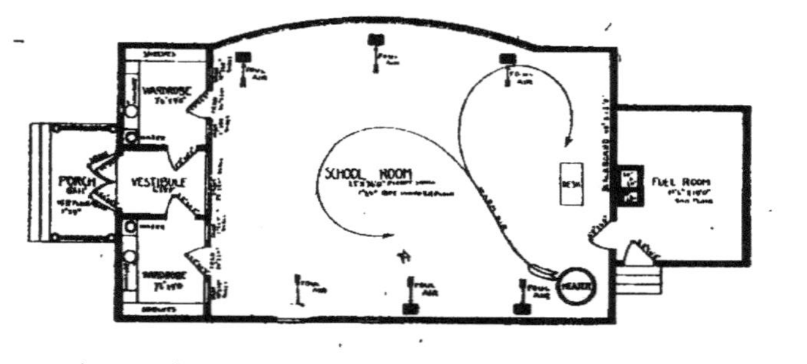

Blue prints consist of floor plan; front and side elevations; foundation plan; porch detail. Complete typewritten specifications with each set of plans.

"The Essex"

Price of Plans and Specifications

$5.00

Full and complete working plans and specifications of this house will be furnished for $5.00. Cost of this house is from $2,100 to $2,200, according to the locality in which it is built.

Floor Plans of "The Essex"

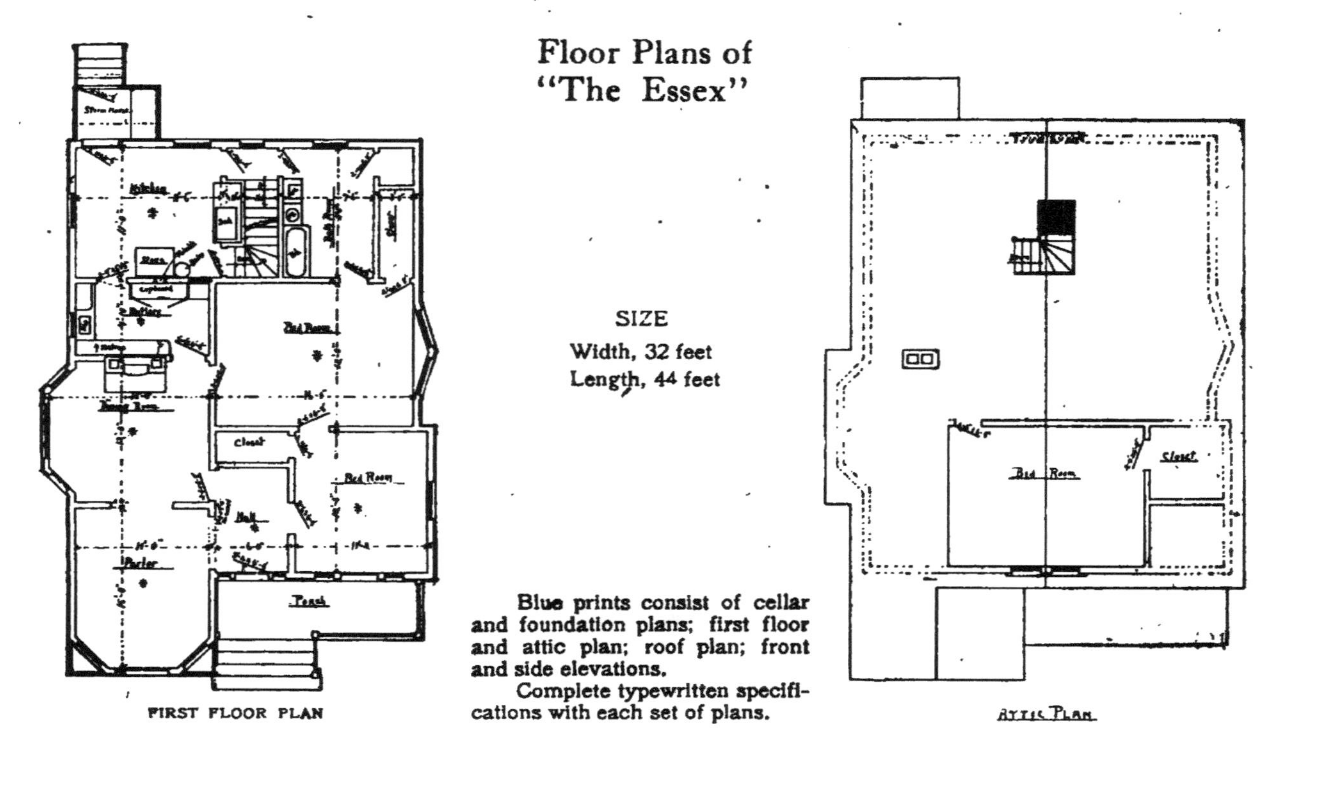

FIRST FLOOR PLAN

ATTIC PLAN

SIZE
Width, 32 feet
Length, 44 feet

Blue prints consist of cellar and foundation plans; first floor and attic plan; roof plan; front and side elevations.

Complete typewritten specifications with each set of plans.

"The Omaha"

Price of Plans and Specifications

$5.00

Full and complete working plans and specifications of this house will be furnished for $5.00. Cost of this house is from $1.400 to $1,500, according to the locality in which it is built.

Floor Plans of "The Omaha"

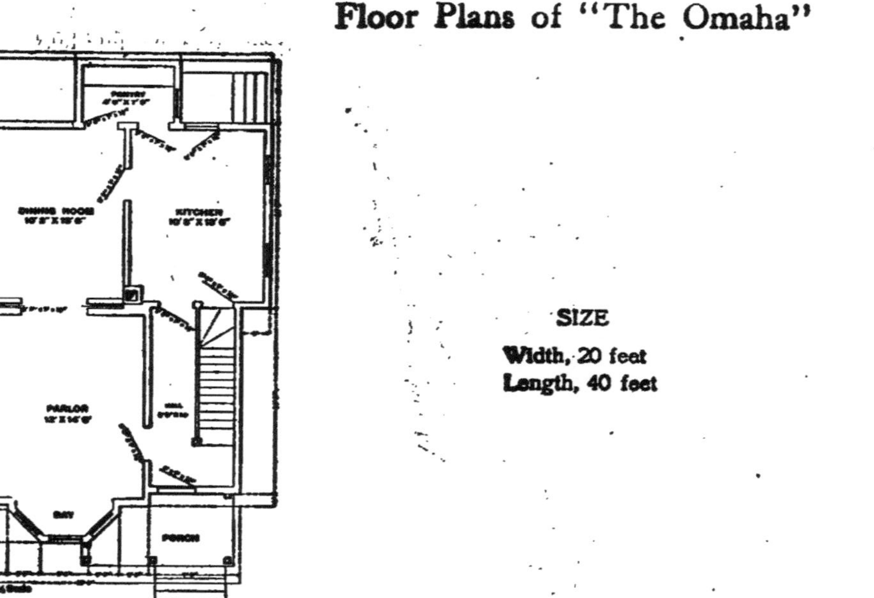

SIZE

Width, 20 feet
Length, 40 feet

Blue prints consist of cellar and foundation plan; floor plans; roof plan; front and side elevations. Complete typewritten specifications with each set of plans.

"St. Japes Church"

Price of Plans and Specifications

$10.00

Full and complete working plans of this church will be furnished for $10.00.
This church has been erected at a cost of $8,500.

Floor Plan of "St. James Church"

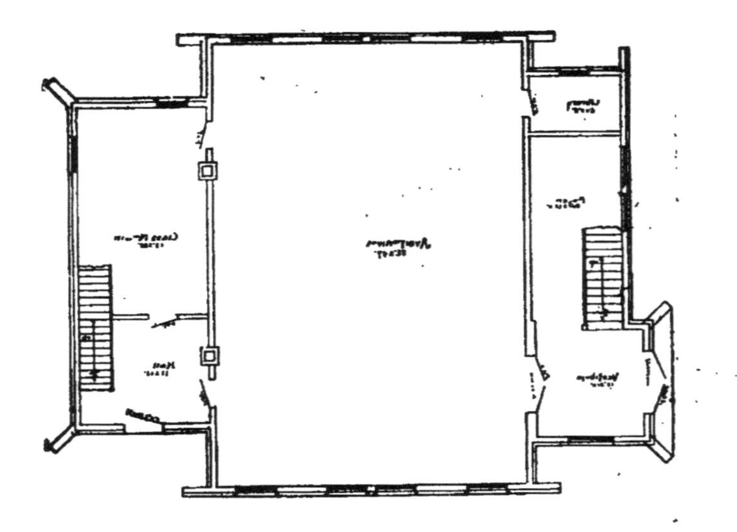

Blue prints consist of floor plan; front and side elevations.

Complete typewritten specifications with each set of plans.

"The Flora"

Price of Plans and Specifications

$5.00

Full and complete working plans and specifications of this house will be furnished for $5.00.

Cost of house is from $1,050 to $1,150, according to the locality in which it is built.

Floor Plan of "The Flora"

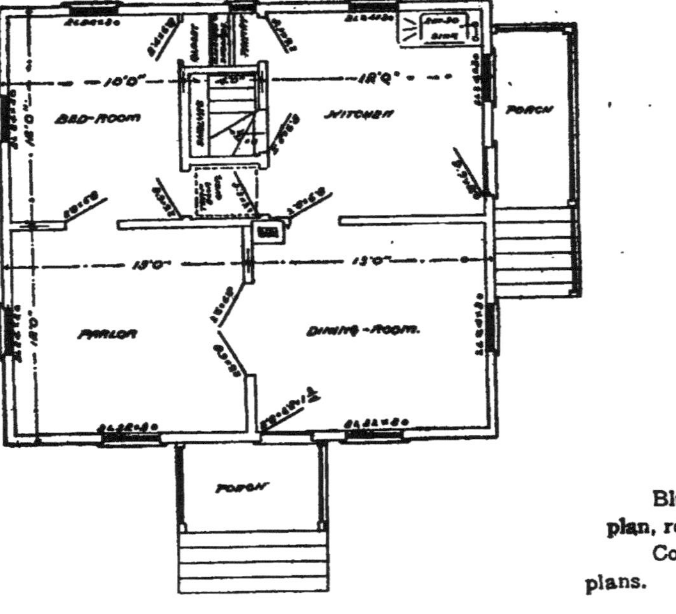

SIZE

Width, 24 feet
Length, 26 feet

Blue prints consist of cellar and foundation plan; floor plan, roof plan; front and side elevation.

Complete typewritten specifications with each set of plans.

"The Collingwood"

Price of Plans and Specifications

$6.00

Full and complete working plans and specifications of this house will be furnished for $6.00. Cost of this house is from $3,100 to $3,200, according to the locality in which it is built.

Floor Plans of "The Collingwood"

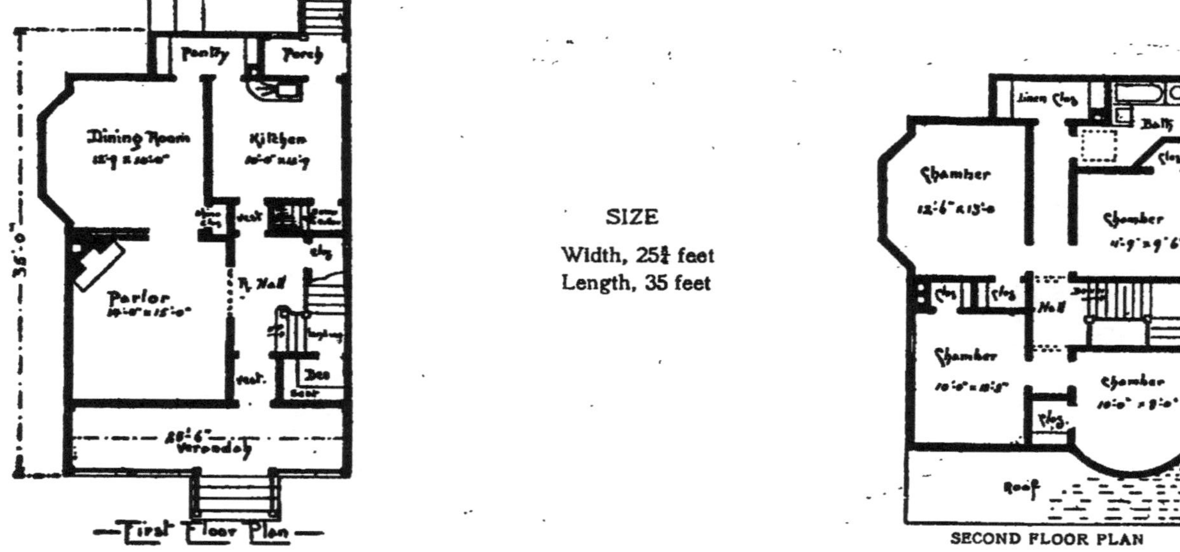

SIZE

Width, 25¾ feet
Length, 35 feet

Blue prints consist of cellar and foundation plan; roof plan; floor plans; front and side elevations.
Complete typewritten specifications with each set of plans.

"The Clover"

Price of Plans and Specifications

$5.00

Full and complete working plans and specifications of this house will be furnished for $5.00. Cost of this house is from $1,000 to $1,100, according to the locality in which it is built.

Floor Plan of "The Clover"

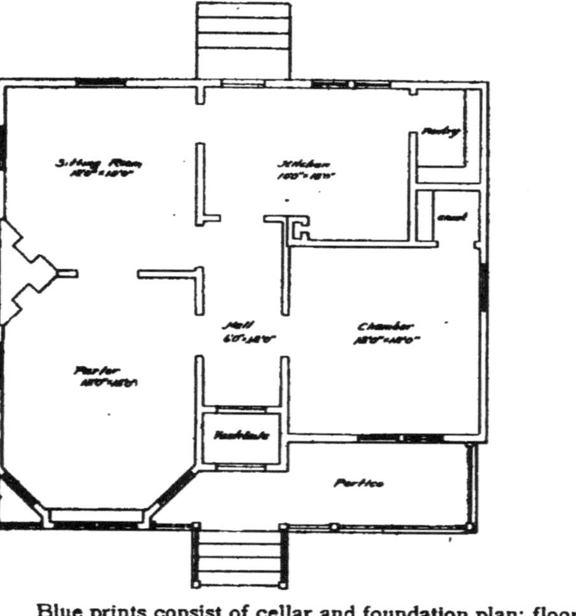

SIZE

Width, 24 feet   Length, 28 feet

Blue prints consist of cellar and foundation plan; floor plan; roof plan; front and side elevation. Complete typewritten specifications with each set of plans.

"The Foster"

Price of Plans and Specifications

$6.00

Full and complete working plans and specifications of this house will be furnished for $6.00.
Cost of this house is from $3,000 to $3,100, according to the locality in which it is built.

Floor Plans of "The Foster"

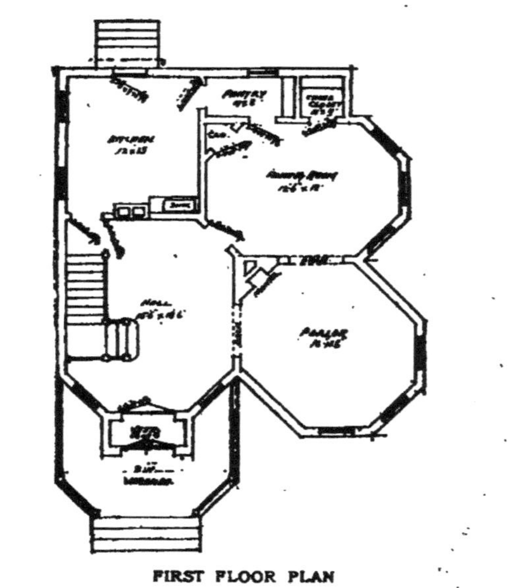

FIRST FLOOR PLAN

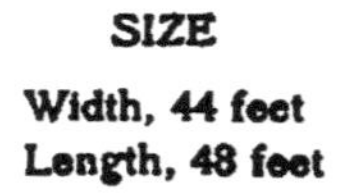

SIZE

Width, 44 feet
Length, 48 feet

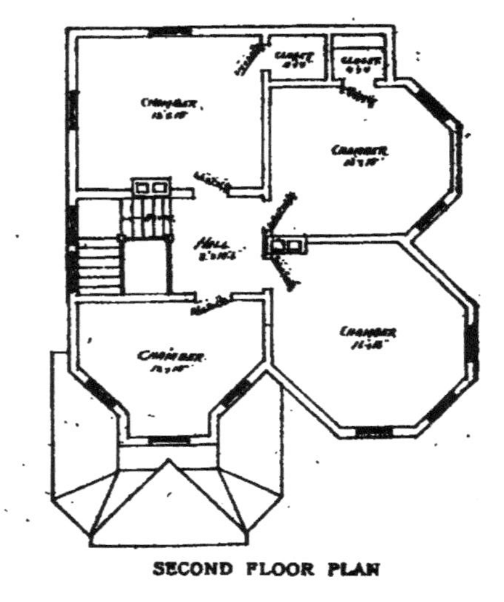

SECOND FLOOR PLAN

Blue prints consist of cellar and foundation plan; floor plans; roof plan; front and side elevations. Complete typewritten specifications with each set of plans.

"The Logan"

Price of Plans and Specifications

$7.00

Full and complete plans and specifications of this house will be furnished for $7.00.
Cost of this house is about $4,200.

Floor Plans of "The Logan"

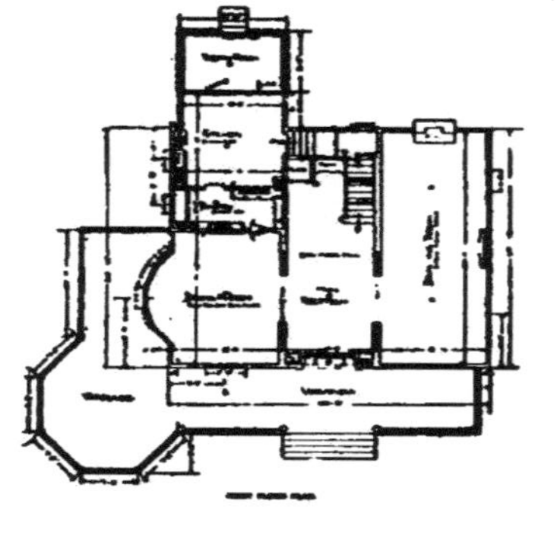

SIZE:

Length, 60 feet
Width, 61 feet

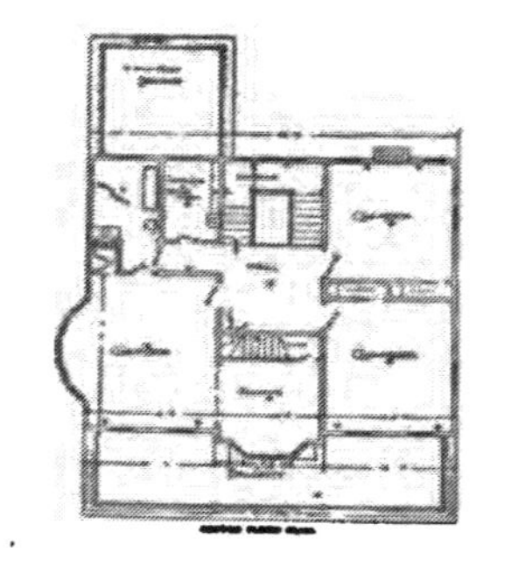

Blue prints consist of cellar and foundation plan; roof plan; floor plans; front and side elevations. Complete typewritten specifications with each set of plans.

"The Hicks"

Price of Plans and Specifications

$5.00

Full and complete working plans and specifications of this house will be furnished for $5.00. Cost of this house is from $1,200 to $1,300, according to the locality in which it is built.

Floor Plan of "The Hicks"

SIZE

Width, 24 feet   Length, 40 feet

Blue prints consist of cellar and foundation plan; floor plan; roor plan; front and side elevation.
Complete typewritten specifications with each set of plans.

"The Kimball"

Price of Plans and Specifications

$5.00

Full and complete working plans and specifications of this house will be furnished for $5.00. Cost of this house is from $1,000 to $1,200, according to the locality in which it is built.

Floor Plan of "The Kimball"

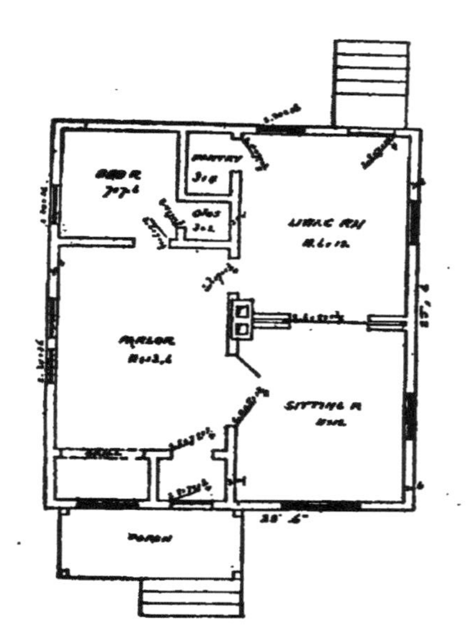

SIZE

Width, 24 feet
Length, 28 feet

Blue prints consist of cellar and foundation plan; floor plan; roof plan; front and side elevations.
Complete typewritten specifications with each set of plans.

"The Glen Flora"

Price of Plans and Specifications

$5.00

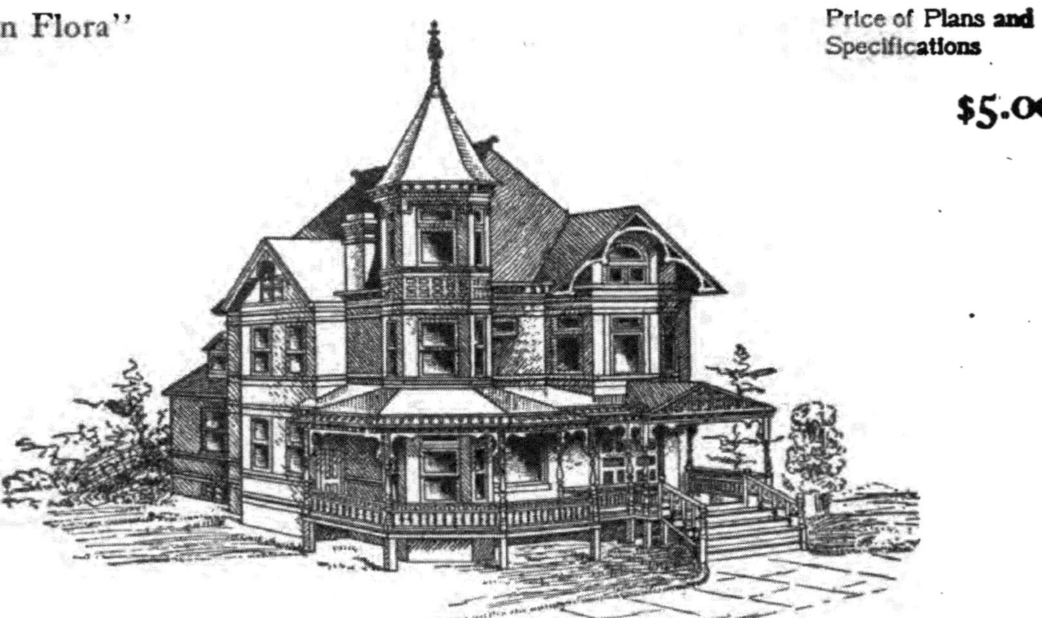

Full and complete working plans and specifications of this house will be furnished for $5.00
Cost of this house is from $2,300 to $2,500, according to the locality in which it is built.

Floor Plans of "The Glen Flora"

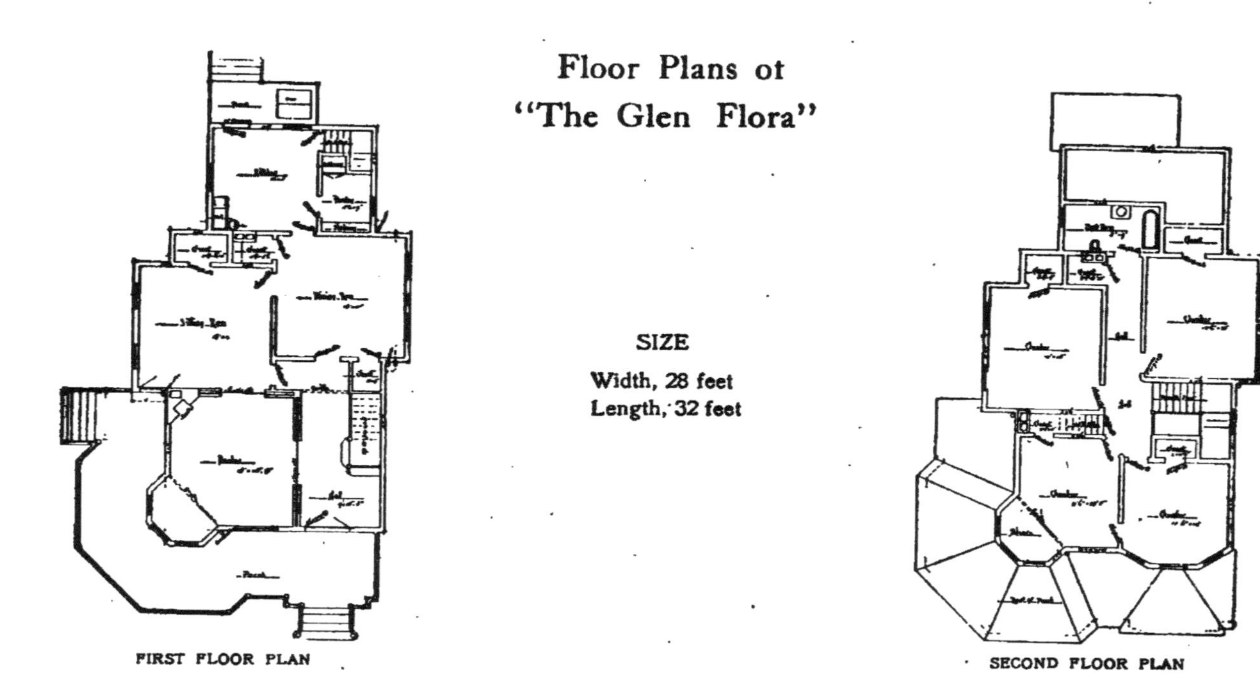

FIRST FLOOR PLAN

SECOND FLOOR PLAN

Blue prints consist of cellar and foundation plan; floor plans; front and side elevations. Complete typewritten specifications with each set of plans.

"The Brookdale"

Price of Plans and Specifications

$5.00

Full and complete working plans and specifications of this house will be furnished for $5.00. Cost of this house is from $2,500 to $2,700, according to the locality in which it is built.

Floor Plans of "The Brookdale"

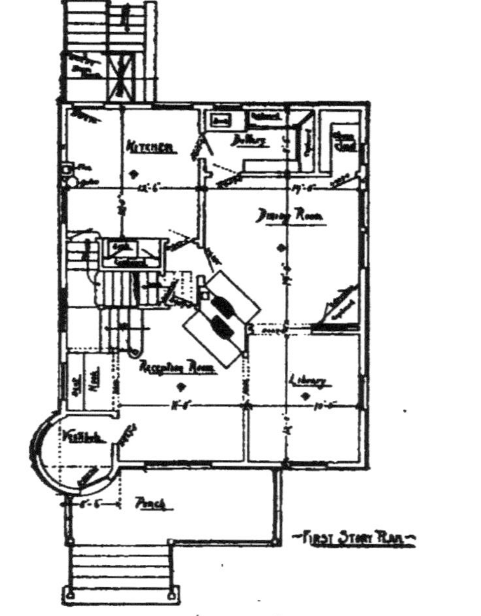

SIZE

Width, 32 feet
Length, 38 feet

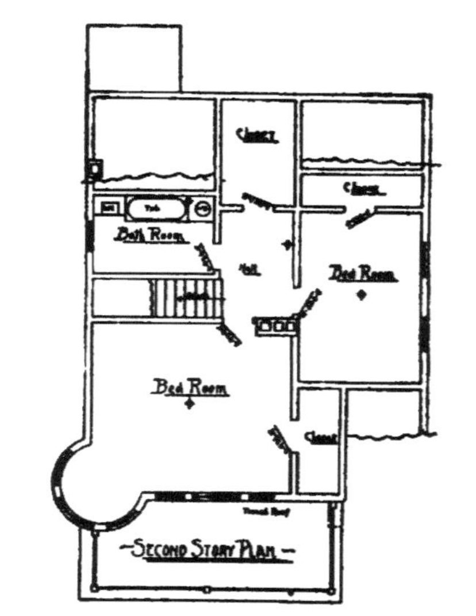

Blue prints consist of cellar and foundation plan; first and second floor plans; roof plan; front and side elevations. Complete typewritten specifications with each set of plans.

"The Russell"

Price of Plans and Specifications

$5.00

Full and complete working plans and specifications of this house will be furnished for $5.00. Cost of this house is from $1,200 to $1,400, according to the locality in which it is built.

Floor Plans of "The Russell"

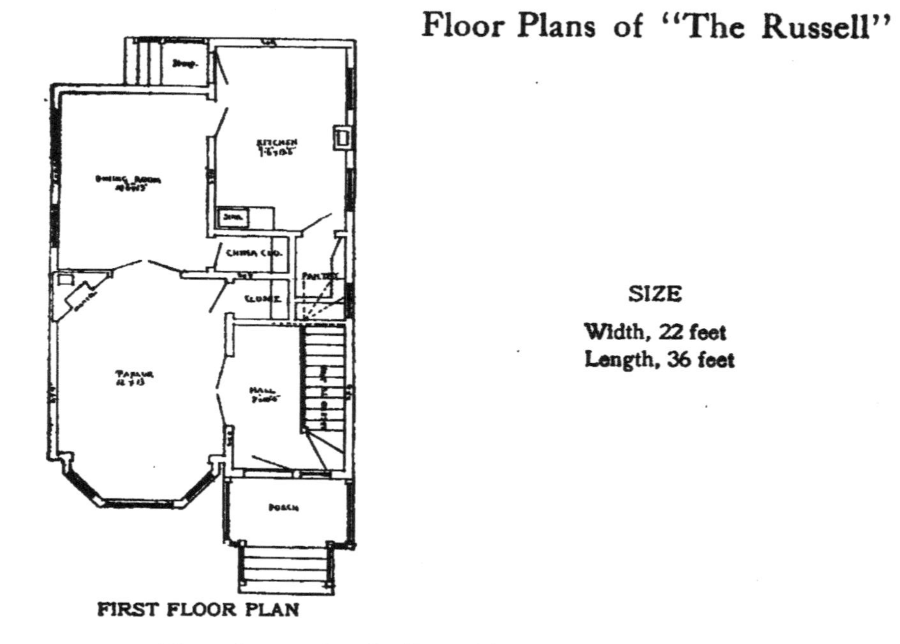

FIRST FLOOR PLAN

SIZE

Width, 22 feet
Length, 36 feet

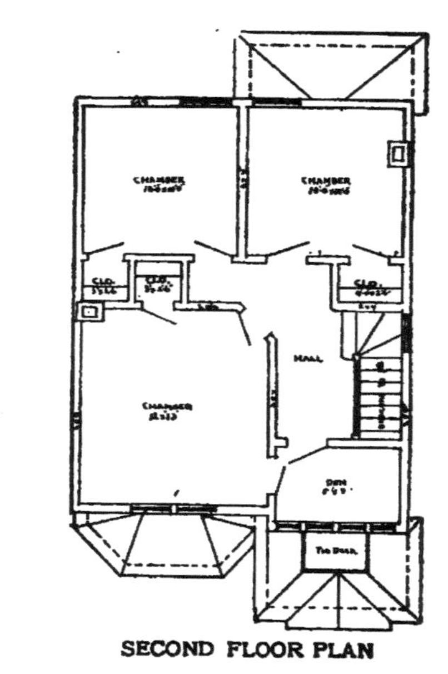

SECOND FLOOR PLAN

Blue prints consist of cellar and foundation plan; floor plans; front and side elevations.
Complete typewritten specifications with each set of plans.

"The King"

Price of Plans and Specifications

$5.00

Full and complete plans and specifications of this house will be furnished for $5.00. Cost of this house is about $1,200.

Floor Plan of "The King"

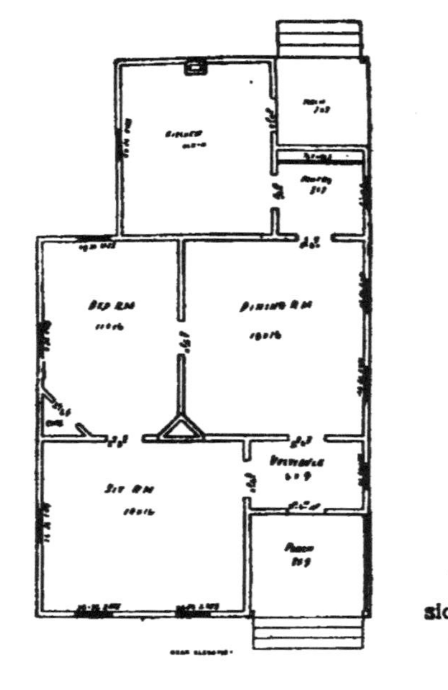

SIZE:
Length, 26 feet
Width, 44 feet

Blue prints consist of cellar and foundation plan; roof plan; floor plans; front and side elevations.

Complete typewritten specifications with each set of plans.

"The Durstin"

Price of Plans and Specifications

$5.00

Full and complete plans and specifications of this house will be furnished for $5.00.
Cost of this house is from $1,600 to $1,700, according to the locality in which it is built.

Floor Plans of "The Durstin"

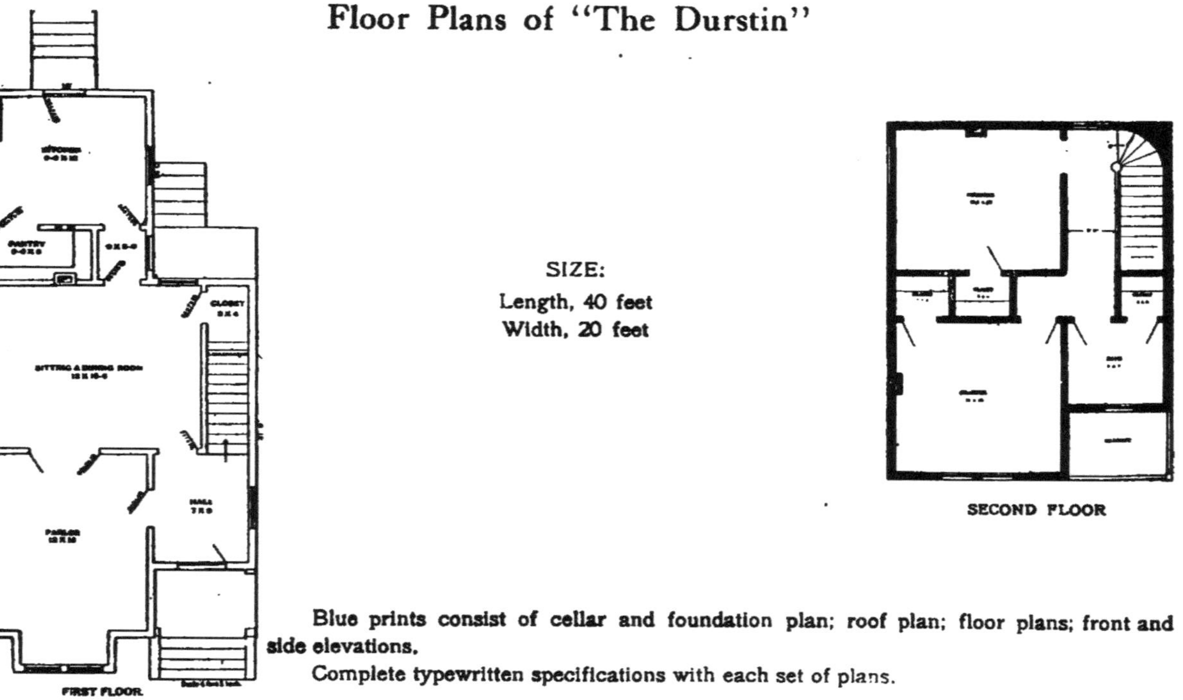

SIZE:
Length, 40 feet
Width, 20 feet

Blue prints consist of cellar and foundation plan; roof plan; floor plans; front and side elevations.

Complete typewritten specifications with each set of plans.

"The Riverside"

Price of Plans and Specifications

$7.00

Full and complete plans and specifications of this house will be furnished for $5.00.
Cost of this house is from $4,200 to $4,300, according to the locality in which it is built.

Floor Plans of "The Riverside"

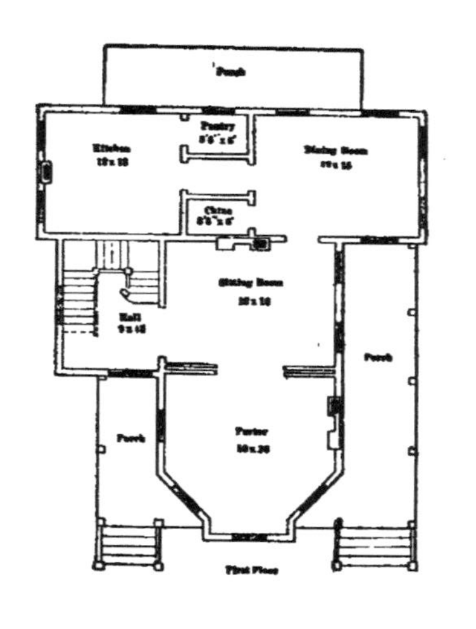

SIZE:

Length, 42 feet
Width, 35 feet

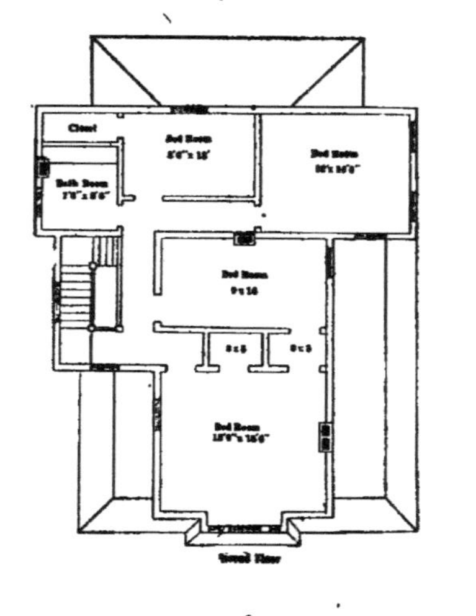

Blue prints consist of cellar and foundation plan; roof plan; floor plans; front and side elevations.
Complete typewritten specifications with each set of plans.

"The Parvins"

Price of Plans and Specifications

$7.50

Full and complete plans and specifications of this house will be furnished for $5.00. Cost of this house is $7,000.

Floor Plans of "The Parvins"

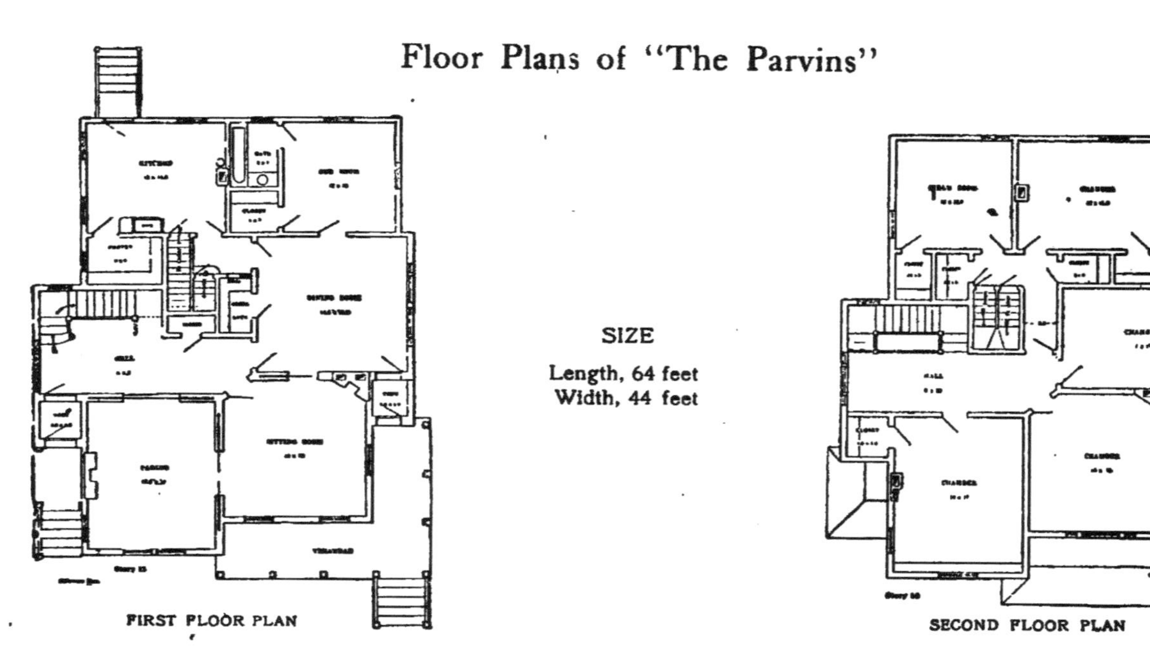

SIZE

Length, 64 feet
Width, 44 feet

Blue prints consist of cellar and foundation plan; roof plan; floor plans; front and side elevations. Complete typewritten specifications with each set of plans.

"The Corbin"

Price of Plans and Specifications

$5.00

Full and complete plans and specifications of this house will be furnished for $5.00.
Cost of this house is from $2,600 to $2,700, according to the locality in which it is built.

Floor Plans of "The Corbin"

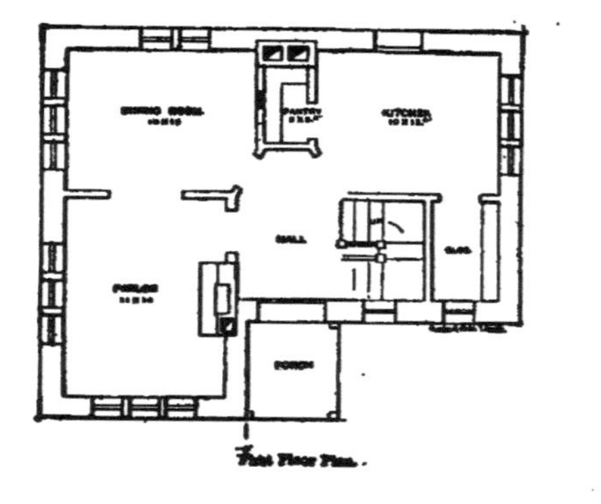

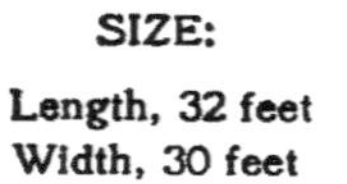

SIZE:

Length, 32 feet
Width, 30 feet

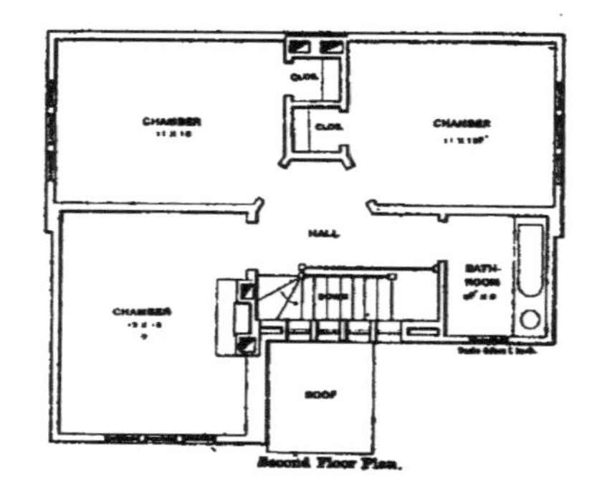

Blue prints consist of cellar and foundation plan; roof plan; floor plans; front and side elevations. Complete typewritten specifications with each set of plans.

"The Bennett"

Price of Plans and Specifications

$5.00

Full and complete plans and specifications of this house will be furnished for $5.00.
Cost of this house is from $2,200 to $2,300, according to the locality in which it is built.

Floor Plans of "The Bennett"

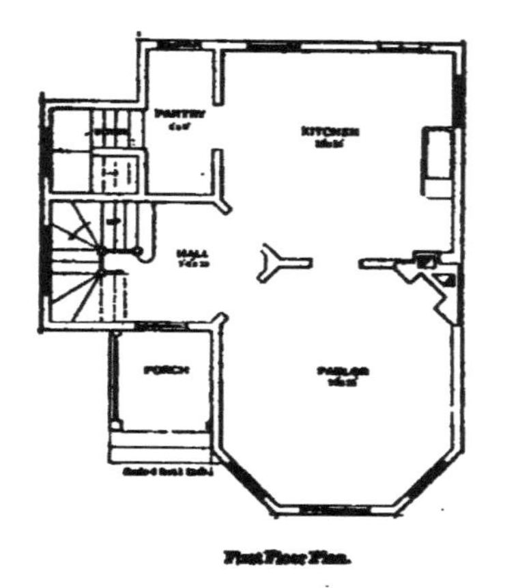

First Floor Plan.

SIZE;

Length, 30 feet
Width, 24 feet

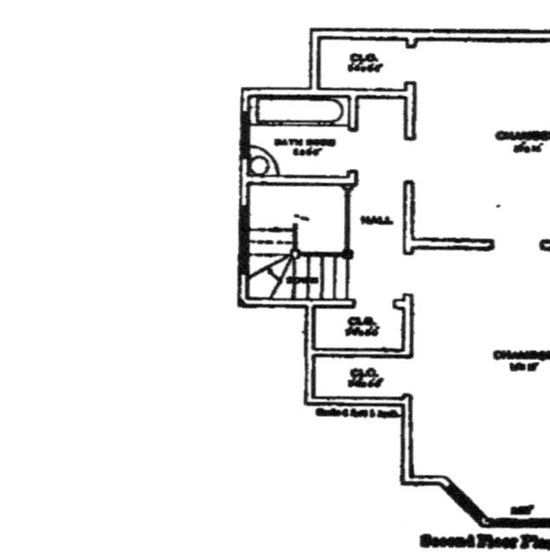

Second Floor Plan.

Blue prints consist of cellar and foundation plan; roof plan; floor plans; front and side elevations. Complete typewritten specifications with each set of plans.

"The West Baden"

Price of Plans and Specifications

$6.00

Full and complete working plans and specifications of this house will be furnished for $6.00. Cost of this house is from $3,000 to $3,100, according to the locality in which it is built.

Floor Plans of "The West Baden"

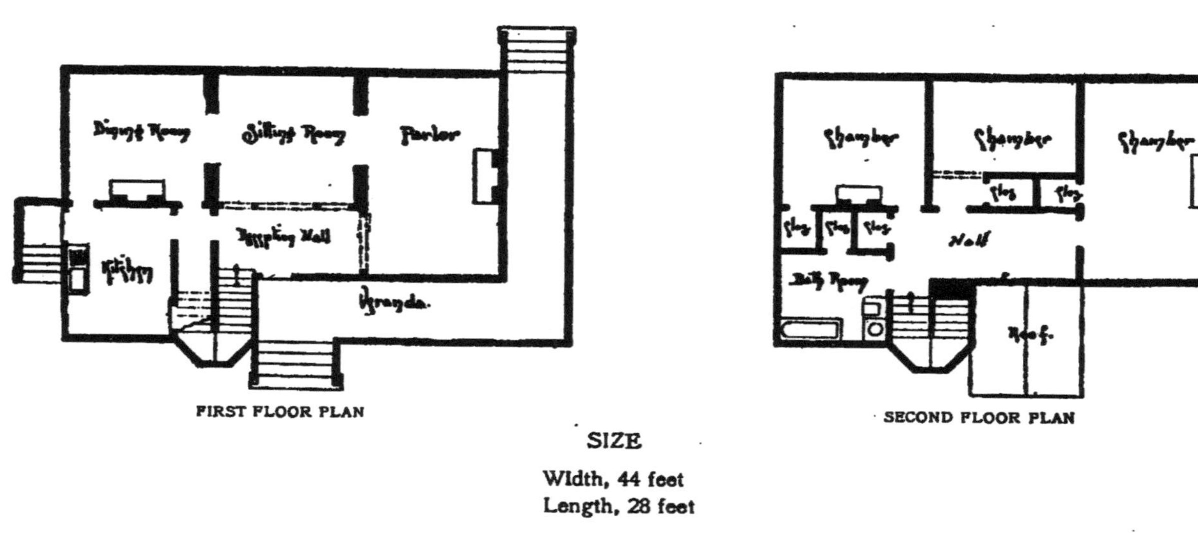

FIRST FLOOR PLAN

SECOND FLOOR PLAN

SIZE

Width, 44 feet
Length, 28 feet

Blue prints consist of cellar and foundation plan; floor plans; roof plan; front and side elevations. Complete typewritten specifications with each set of plans.

"The Oregon"

Price of Plans and Specifications

$5.00

Full and complete working plans and specifications of this house will be furnished for $5.00. Cost of this house is from $1,500 to $1,600, according to the locality in which it is built.

Floor Plan of "The Oregon"

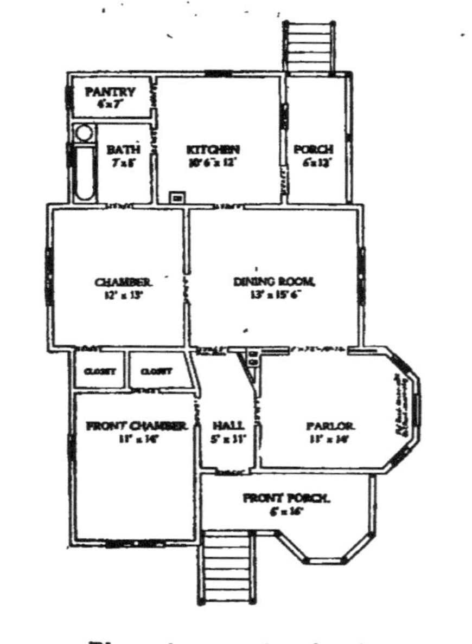

SIZE

Width, 34 feel

Length, 48 feet

Blue prints consist of cellar and foundation plan; floor plan; roof plan; front and side elevations. Complete typewritten specifications with each set of plans.

"The Orchard Crest"

Price of Plans and Specifications

$6.00

Full and complete working plans and specifications of this house will be furnished for $6.00. Cost of this house is from $3,000 to $3,200, according to the locality in which it is built.

Floor Plans of "The Orchard Crest"

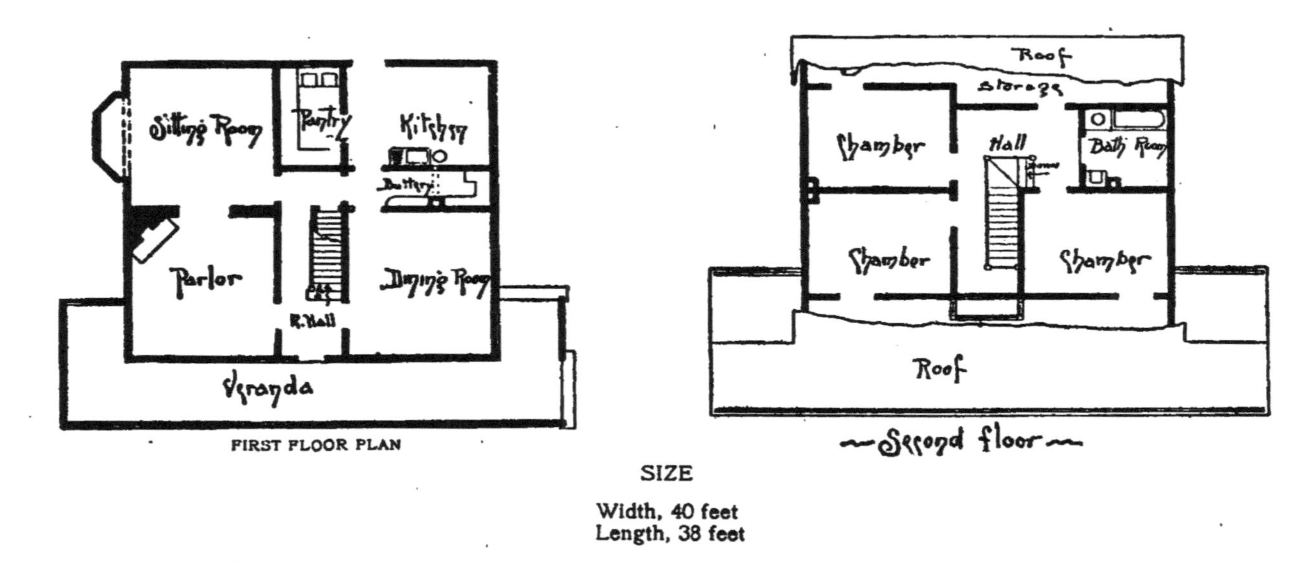

FIRST FLOOR PLAN

—Second floor—

SIZE

Width, 40 feet
Length, 38 feet

Blue prints consist of cellar and foundation plan; roof plan; floor plans; front and side elevations. Complete typewritten specifications with each set of plans.

"The Marshall"

Price of Plans and Specifications

$5.00

Full and complete working plans and specifications of this house will be furnished for $5.00.
Cost of this house is from $750 to $800, according to the locality in which it is built.

Floor Plan of "The Marshall"

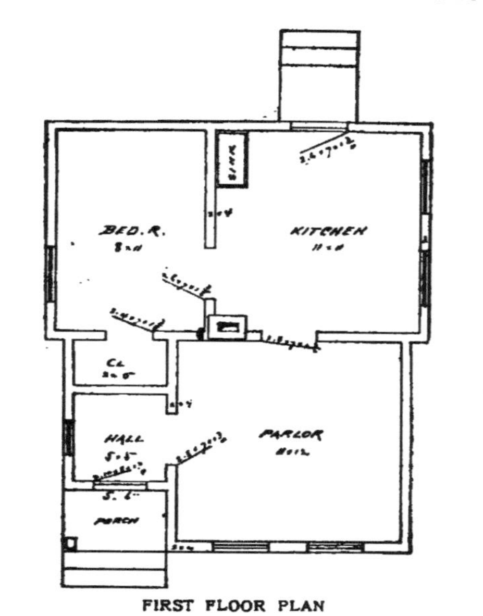

FIRST FLOOR PLAN

SIZE

Width, 20 feet
Length, 24 feet

Blue prints consist of cellar and foundation plan; roof plan; floor plan; front and side elevations. Complete typewritten specifications with each set of plans.

"The Tacoma"

Price of Plans and Specifications

$5.00

Full and complete working plans and specifications of this house will be furnished for $5.00.
Cost of this house is from $1,550 to $1,600, according to the locality in which it is built.

Floor Plans of "The Tacoma"

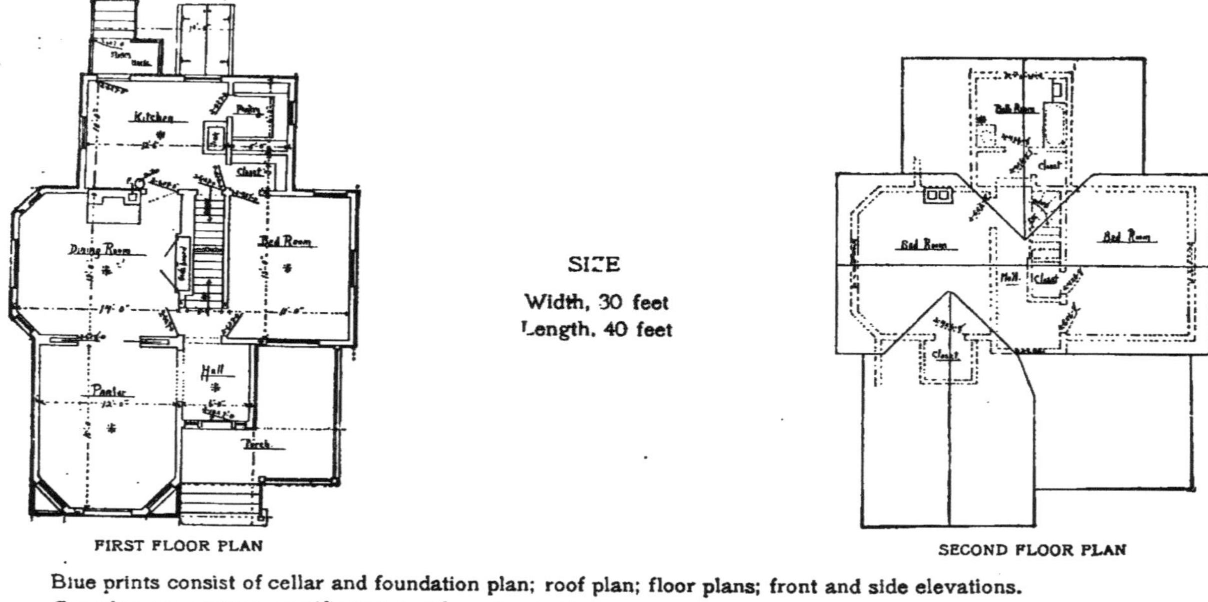

Blue prints consist of cellar and foundation plan; roof plan; floor plans; front and side elevations.
Complete typewritten specifications with each set of plans.

"The San Jose"

Price of Plans and Specifications

$5.00

Full and complete working plans and specifications of this house will be furnished for $5.00. Cost of this house is from $1,000 to $1,100, according to the locality in which it is built.

Floor Plan of "The San Jose"

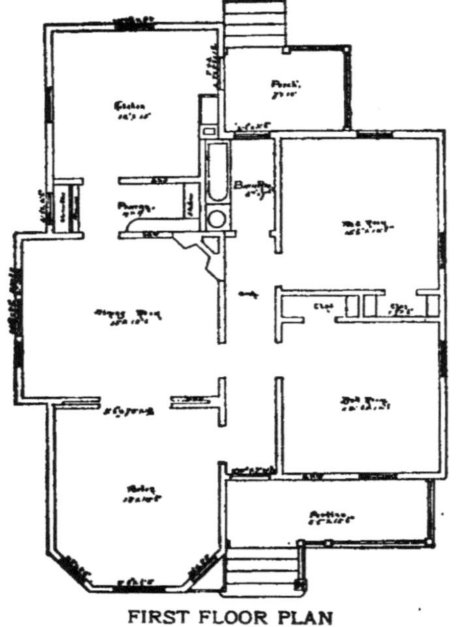

FIRST FLOOR PLAN

SIZE

Width, 32 feet   Length, 46 feet

Blue prints consist of cellar and foundation plan; floor plan; roof plan; front and side elevation.

Complete typewritten specifications with each set of plans.

Modern Carpentry

A PRACTICAL MANUAL

FOR CARPENTERS AND WOOD WORKERS GENERALLY

y FRED T. HODGSON, Architect, Editor of the National Builder, Practical Carpentry, Steel Square and Its Uses, etc., etc.

A NEW, complete guide, containing **hundreds of quick methods** for performing work in **carpentry, joining and general wood-work.** Like all of Mr. Hodgson's works, it is written in a simple, every-day style, and does not bewilder the working-man with long mathematical formulas or abstract theories. The illustrations, of which there are many, are explanatory, so that any one who can read plain English will be able to understand them easily and to follow the work in hand without difficulty.

The book contains methods of **laying roofs, rafters, stairs, floors, hoppers, bevels, joining mouldings, mitering, coping, plain hand-railing, circular work, splayed work,** and many other things the carpenter wants to know to help him in his every day vocation. It is the **most complete** and **very latest** work published, being **thorough, practical** and **reliable.** One which no carpenter can afford to be without.

The work is printed from new, large type plates on a superior quality of cream wove paper, durably bound in English cloth.

Price - - - - - $1.00

FREDERICK J. DRAKE & CO.

211-213 E. Madison St., Chicago.

Lightning Source UK Ltd.
Milton Keynes UK
UKHW020702020122
396454UK00006B/1349